Integrated Science

Volume 21

Editor-in-Chief

Nima Rezaei, Tehran University of Medical Sciences, Tehran, Iran

The **Integrated Science** Series aims to publish the most relevant and novel research in all areas of Formal Sciences, Physical and Chemical Sciences, Biological Sciences, Medical Sciences, and Social Sciences. We are especially focused on the research involving the integration of two of more academic fields offering an innovative view, which is one of the main focuses of Universal Scientific Education and Research Network (USERN), science without borders.

Integrated Science is committed to upholding the integrity of the scientific record and will follow the Committee on Publication Ethics (COPE) guidelines on how to deal with potential acts of misconduct and correcting the literature.

Adela Cortina · Mauricio Correa Casanova
Editors

Applied Ethics from Medicine to Humor

EDICIONES UC

Springer

Editors
Adela Cortina
Department of Philosophy and Moral and Political Philosophy
University of Valencia
Valencia, Spain

Mauricio Correa Casanova
Institute of Philosophy Institute of Applied Ethics
Pontifical Catholic University of Chile
Santiago, Chile

ISSN 2662-9461 ISSN 2662-947X (electronic)
Integrated Science
ISBN 978-3-031-59008-5 ISBN 978-3-031-59009-2 (eBook)
https://doi.org/10.1007/978-3-031-59009-2

Jointly published with EdicionesUC
ISBN of the Co-Publisher's edition: 978-9-561-42501-9

Translation from the Spanish language edition: "Etica aplicada desde la medicina hasta el humor" by Adela Cortina and Mauricio Correa Casanova, © Ediciones UC 2019. Published by Ediciones UC. All Rights Reserved.

The original submitted manuscript has been translated into English. The translation was done using artificial intelligence. A subsequent revision was performed by the author(s) to further refine the work and to ensure that the translation is appropriate concerning content and scientific correctness. It may, however, read stylistically different from a conventional translation.

This Springer imprint is published by the registered company Springer Nature Switzerland AG
The registered company address is: Gewerbestrasse 11, 6330 Cham, Switzerland

Foreword

What is known as "applied ethics" began its splendorous journey in the 1970s. Since then, the expression has been used profusely, first in the USA and then worldwide. However, reactions to it have varied greatly.

On the one hand, professionals, citizens, and governments have every reason to demand a higher ethical level in the various social spheres and, above all, to demand the institutionalization of this level in committees, commissions, guidelines, and well-defined documents so that it is the social reality *itself* that requires moral philosophy to commit itself to everyday life.

On the other hand, the response from moral philosophers has been quite diverse. While authors as far apart as Jürgen Habermas and Alasdair MacIntyre refuse to accept that ethics contains an applied dimension (the former because he believes that this is the task of the social sciences or education, the latter because he considers applied ethics to be based on an error), a good number of neo-Aristotelians and deontologists, as well as utilitarians, from the outset have taken the path of application. Thus were born applied ethics.

Initially, there were land ethics, development ethics, bioethics, and company (or business) ethics. Later, professional ethics (genetics, ecoethics, media ethics, political ethics, infoethics, consumer ethics, and so many others) emerged, covering different spheres of social life—so much so that it can be said that applied ethics are already a social reality, irreversible in the different countries, giving substance to the Hegelian dream of incarnating morality in institutions. The task of practical reason is not only to enunciate what should be done but also to take flesh in institutions, transforming them from within. Hence, applied ethics is an inescapable element of any institutional design that wants to function well and even of any research project that wants to be approved.

Periodicals and bibliographies on applied ethics have proliferated unprecedentedly in different fields. Companies, public administrations, and professional associations have incorporated ethical ways of knowing and doing, while universities all over the world have created different chairs, degrees, and research centers devoted to the application of ethical theory in a variety of fields.

This volume brings together a series of papers that address different areas of applied ethics. By now, some of them are well known for their wide dissemination, such as biomedicine, development, ecology, economics, business, and professions. Others are less known and their novelty presents unprecedented challenges in fields such as neuroethics, sports, city, and humor. The main objective of each chapter is to offer teachers and students, professionals in various fields, members of civic organizations, and all those interested in these subjects a review of ethical perspectives that contribute to incarnating morality in their respective institutions, which is the hinge of applied ethics.

Valencia, España Adela Cortina
Santiago, Chile Mauricio Correa Casanova

Contents

Biomedical Ethics

1

Juan Pablo Faúndez Allier

Abstract

The scenarios that follow various justifications of deliberation methods allow us to contrast different criteria application strategies that explicitly show how this branch of applied ethics, in the end, does nothing more than reflect procedural models sustained in a fundamental ethical interpretation. In this sense, biomedical ethics has a special relevance, advancing in the dialogic positioning of principles and values that today serve not only this area of applied ethics. Debate within this area of ethics, which has provided the classic framework of principles and subsequently of values used by different applied sub-currents, is fully alive and in dialogue with the societies that constitute the changing moment that we are opening as a culture.

Keywords

Biomedical ethics • Methodology • Deliberation • Principles • Values • Clinic

1 Towards a Terminological Retrospection

Mentioning the term "biomedical ethics" leads immediately to the remarkable work published by Beauchamp and Childress in 1979, which closed the journey, now with "non-maleficence," of the famous principles that *The Belmont Report* had pointed out as guarantors of ethical reflection on the treatment of life in the contemporary West. The work of the American authors was rightly entitled *Principles*

J. P. Faúndez Allier (✉)
Pontificia Universidad Católica de Valparaíso, Valparaiso, Chile
e-mail: juan.faundez@pucv.cl

A. Cortina and M. Correa Casanova (eds.), *Applied Ethics from Medicine to Humor*, Integrated Science 21, https://doi.org/10.1007/978-3-031-59009-2_1

of Biomedical Ethics [4], and with it was mentioned the research, experimentation and application of ethical criteria in situations related to health sciences in the stages of origin, maintenance and termination of life [1]. In this way, the principles of respect for persons ("autonomy"), "beneficence," and "justice," established by the landmark report of the *National Commission for the Protection of Human Subjects of Biomedical and Behavioral Research*, commissioned by the United States Congress in 1974 and concluded in 1978, were completed by the necessary differentiation made by Beauchamp and Childress when distinguishing that "beneficent" actions did not always coincide with "non-maleficent" ones.

But this matter of analysis had already been put forward in 1970 and 1971, when the American oncologist Van Rensselaer Potter exposed in the article "Bioethics: Science of Survival" [26] and later deepened in his book *Bioethics: Bridge to the Future* [27] that, precisely, the concern about the approach that should be made from ethics to "bios" in medicine was not only not redundant, but should be elucidated and deepened with the prolific research and medical technological application that began to take the human being to crossroads and situations as never before. The study of complex life, and its conditions of possibility and permanence had to be approached as a proper object of analysis in interpersonal relations, and the diverse links that revolve around this phenomenon: doctor-user; doctor-health system; health system-user; and in the reciprocal effects that were initiated. Hence, Potter made explicit the broader term "bioethics," understood as the concept that in some way we could recognize as the generic continent and richer in nuances than that of biomedical ethics, since bioethics deals with, as well-defined by Reich [28], the "systematic study of human conduct in the field of life sciences and health care, as such conduct is examined in the light of moral principles and values" (Vol 1, XIX).

In Potter's texts, the mother name of biomedical ethics, bioethics, acknowledged a second paternity of American origin. Its true birth dated back to the year in which Jahr mentioned this term [22] to refer to the awareness he assumed was relative to the effects of ethical implications that could arise from the interaction between human beings and plants and animals. Hence the title of his article: "Bioethik: Eine Umschau über die ethischen Beziehungen des Menschen zu Tier und Pflanze" ("Bio-ethics: A survey of the ethical relationships between humans and animals and plants") [22]. The word used by Jahr, bioethics, was definitely visionary, since it focused on the adequate participation that the human being should assume at the moment of interacting with beings of diverse hierarchies that coexist in the plurality of ecosystems of our planet. Jahr's approach, as a Protestant philosopher and theologian, sustained a critique of the Western vision that assumes a hierarchical leveling that despises beings inferior to humans.

With this retrospective journey, I wanted to show that in the conceptual relationship between bioethics and biomedical ethics it is convenient to make a marriage between the terms, although marking a subtle difference. The second has been used to express an attempt at precision in relation to the broader function that bioethics has assumed over the years. Strictly speaking, biomedical ethics must be distinguished from other branches of bioethics that will take charge of more specific areas that we can now differentiate, and whose contrast did not appear clearly in the early years of its development. In this sense, today, the following are considered as areas of bioethics that are distinguished from biomedical ethics: "health ethics," "clinical ethics," "ethics of human reproduction," "gene-ethics," "eco-ethics," "zoo-ethics," "tanato-ethics" and "neuro-ethics." These are sub-disciplines of bioethics that are advancing in parallel and are increasingly consolidating their own lines of research and conceptual domain [6].

Given that an in-depth study of each of the specific areas of bioethics is beyond the scope of this chapter, and that in the following parts of this book some sub-species of bioethics will be dealt with (according to the task entrusted by the editors), I will focus on developing the installation and development of biomedical ethics, strictly speaking. Having said this, I distinguish from the outset that the mere advancement of this field of research in various cultural areas would also open us to countless comparative details, so I have deemed it appropriate to focus on the Latin American context, in the understanding, moreover, that this is the audience to which this volume was originally addressed. This does not prevent the analysis of the first steps in the formal study of biomedical ethics from necessarily situating us in the North American context, which is where this sub-discipline of bioethics arises.

2 The Beginnings of the Institutional and Normative Organization of Biomedical Ethics

Biomedical ethics has a close relationship with health ethics, which is the subspecialty of bioethics that is responsible for interpreting the interrelationship of institutional bodies that manage the provision of health to the population through social policies, and clinical ethics, which aims to elucidate the criteria for ethical interpretation of the actions that link health personnel with the user of the system. Making a difference, biomedical ethics, considering the described relationships of health and clinical type, focuses on the reflection on the morality of research methods, experimentation and practical application of the findings that the experimental analysis is raising.

In the years in which the neologism "bio-ethik," offered by Jahr, remained practically unnoticed, the joint proposal put forward by the philosopher Daniel Callahan and the psychiatrist Willard Gayling, organizers of several meetings between scientists, philosophers and doctors in the search for solving the moral problems of the performance of medicine from an integrated perspective, arose in 1968 [7]. For this purpose, the Institute of Society, Ethics and the Life Sciences was born in New York, an institution that would later become known as the *Hastings Center*. The work of the team that was formed in this institute would consist of the study and orientation that could be offered to respond to the social expectations that opened up with the advances in medical technology, generating a series of questions related to the most delicate moments of life maintenance, especially its origin and its end. In this sense, it is worth bearing in mind that in the 1960s, contraception, artificial fertilization and resuscitation at the end of life were beginning to spread, among other topics that required careful reflection to provide consistent arguments, from a moral point of view, and not only plausible ones, from a technical perspective. Biomedical ethics, as a branch of applied bioethics, responded directly to these concerns, providing criteria of moral justification related to the research phase that would illuminate subsequent applications. The analysis of these aspects would later help to specifically recognize this branch of bioethics, placing the effects of research, experimentation and the application of results on the horizon of interest to a wider public than scientific circles.

This is how the Hastings Center consolidated itself as a research center that set international guidelines in biomedical ethics, focusing on studies on biotechnologies applied fundamentally to the field of procreation. The parameters established by the methodological scheme proposed by this center would be followed in the different places where research in this field was beginning to be carried out, assuming an interdisciplinary approach that would gradually outline the projection of this type of experimentation.

In this period, with the use of the term *bioethics*, the professor of oncology at the McArdle Laboratory (University of Wisconsin Madison), Van Rensselaer Potter, enunciates the new discipline that from medicine will seek to integrate aspects of biomedical foundation for its development, although raising the image of the wide "bridge" that would be able to connect medicine, biology and ecology, as elements that would help to land ethics in these contexts. Potter's work would take institutional form in The Joseph and Rose Kennedy Institute for the Study of Human Reproduction and Bioethics, directed by André Hellegers. This institute would begin to generate research projects in parallel to the Hastings Center, establishing a structure that would be very relevant for future institutes that would begin to interrelate research, experimentation and application activities in the medical field. With the death of Hellegers in 1979, the Institute was renamed The Kennedy Institute of Ethics and became definitively linked to Georgetown University.

One of the milestones that definitively establishes the framework and the research and application orientations of biomedical ethics is the work on the principles that emerged from the reflection commissioned by the United States Congress to the National Commission for the Protection of Human Subjects of Biomedical and Behavioural Research, in 1974. Given the outline that the discipline was taking and the global scope that it was gradually reaching, it was necessary to define basic ethical principles that governed research, experimental development and the application of clinical research results in human beings. The Commission soon realized that it was very difficult to achieve clear guidelines in this regard on the basis of abstract moral principles aimed at reaching agreement, so it adopted an inductive method that would facilitate the achievement of consensus. In this way, "basic principles" were obtained that would serve to justify the particular ethical precepts and the evaluation of human action.

After a long period of interdisciplinary deliberation from 1974 to 1978, the principles of biomedical ethics were defined, which would become applicable to bioethics, in its various sub-classifications, over the years. Thus, the *Belmont Report* would give account of a methodological approach that was ordered from the three basic principles, along with recommendations to approach case by case the problems posed by biomedicine, in order to make the procedure more operational (*Ethical Principles and Guidelines for the Protection of Human Subjects of Research*[1]). The principles established by this study (*Basic Ethical Principles*) will be the basic justification for carrying out countless ethical prescriptions and evaluations related to biomedical research, experimentation and application procedures. Thus, three principles are identified as being applicable to research and experimentation with human beings: respect for persons (autonomy), beneficence, and justice.

The first of these was circumscribed by the Report from the following perspective: "Respect for persons incorporates at least two ethical convictions: first, that individuals should be treated as autonomous entities and, second, that persons whose autonomy is diminished should be protected" (a. 4); and then making the content of the same more clearly explicit: "To respect autonomy is to give value to the opinions and choices of persons so considered, and to refrain from obstructing their actions, unless these clearly produce harm to others" (a. 4).

The interpretation of autonomy is not understood by the *Belmont Report* as a structural quality of the human being, but as the possibility of carrying out acts of autonomous choice aimed at achieving a greater operative resolution, through knowledge of the cause and without the influence of any type of external coercion. In other words, it is a functional understanding of the term in the Anglo-Saxon line.

Respect for persons, in their autonomy, is accompanied in the Report by the principle of beneficence. In relation to this normative orientation, the document points out that: "In this sense, two rules have been formulated as complementary expressions of the acts of beneficence: (1) do no harm and (2) maximize the possible benefits and minimize the possible risks" [31]. Beneficence is understood, then, as the principle from which the risks and benefits related to biomedical research and experimentation will be reviewed to ensure that the action generates an effective purpose. Let us distinguish here that the first complementary expression of beneficence is the one that the Report does not initially recognize as a separate principle, which had been specified by Beauchamp and Childress, since it is clearly distinguishable the act of *doing no harm* from that of *doing good*, as I will emphasize a little later.

And, finally, the principle of justice, which is described by the Commission as "fairness in distribution [...] [whereby] equals should be treated equally" [31]. The Report thus pointed to distributive justice, expressed in the proper management

[1] Volumes of annexes containing the analyses of experts who participated in the *National Commission* may be obtained from: Superintendent of Documents, U. S. Government Printing Office, Washington D. C. 20402, under the titles DHEW Publication No. (OS) 780013 and No. (OS) 78-0014.

by exercising an equitable selection of subjects to be subjected to some kind of clinical research or experimentation, as well as the fair distribution of the fruits of the works that were emerging in this line.

Not by chance, a year after the presentation and socialization of the *Belmont Report*, Beauchamp and Childress published *Principles of Biomedical Ethics* [4], a work that emphasizes the importance of the application of principles as guiding criteria for research activities, experimentation and application of results obtained in medical-ethical analysis processes. Going deeper into the study of the National Commission, the authors incorporate a fourth principle which, together with the three previous ones, would come to configure the so-called "Georgetown Mantra": non-maleficence [4]. Non-maleficence differs from the principle of beneficence, since it distinguishes between the obligation not to do evil to others and the obligation to do good to them, the former being a possibility that is granted with more imposing force than the latter, which had not been addressed by the Belmont Report. In the field of biomedical ethics, this distinction is particularly valuable in the area of experimentation and the application of research criteria, which must not only respond to beneficence, but also make explicit the non-maleficence that must be declared before the acceptance of the procedures is requested from those who are willing to agree to them, through the specific adhesion to the intervention that is expressed in the informed consent that the user must give.

The four iconic principles of biomedical ethics (since they were required for research, experimentation and application of procedures) were clearly oriented to the resolution of conflicts that could be projected to the field of clinical ethics, so it was essential to establish the priority criteria linked to their use and the possibility of applying exceptions. The partnership between the authors of *Principles of Biomedical Ethics*, Beauchamp (utilitarian philosopher) and Childress (deontologist theologian), made it difficult to think of a conciliation of arguments when facing conflicts: in the first case, because utilitarians consider that rules are binding to the extent that they are useful, making exceptions justifiable; in the second, because it is considered that principles operate in an absolute way, without giving place to exceptions. This apparent conflict, however, was overcome with the distinction that Davis Ross proposed in 1930 between *prima facie duties* (at-first-sight duties) and *actual duties* (real and effective duties), in *The Right and the Good* [29], being the criterion assumed by medical ethical reflection. In this important book Ross argues that duties must be considered in the abstract as *prima facie* obligatory, without exceptions, so they must be respected when they do not conflict with each other; however, if they are in conflict, it is necessary to prioritize one over the others based on the binding force of the duties in concrete ([5]: 12 ff.). Thus, while the duties are understood *prima facie* from the deontological point of view, in a teleological perspective they are understood as real and effective, which is what happens when their practical application becomes necessarily grounded in experience.

Beauchamp and Childress applied the denomination *prima facie duties* as obligations that are at the same level [14], following a different approach to that which the European schools would assume with respect to the treatment of the principles of biomedical ethics, Specifically, as we will appreciate in the study of how the Hispanic context was deepened in relation to the use of the principles of biomedical ethics, the four principles will be treated, in this case, by means of a justified system of hierarchization. This will highlight criteria of distinction that are not at the service of a mere pragmatic orientation, but are supported by metaphysical foundations, in a line closer to the processes of continental European ethical reflection.

Unlike the Anglo-American model, based on pragmatism, casuism and proceduralism [11], the explanatory foundation for continental European bioethics has gone after a transcendental basis that opens in two different ways: in the case of Germany, seeking this transcendentality through intersubjectivity and intercommunication [21]; and in that of the Mediterranean sphere, affirming that there is an objective reality that constitutes the foundation of human morality and that it is possible to perceive intentionally through processes of valuation. With this, the continental reasoning is opposed to the Anglo-Saxon theoretical-analytical visions that point to reason as a mere subjective instrument of the ethical discourse with which the latter pretend to prevent the achievement of a foundation on objective values that are established in a rational way. The non-cognitive analytical position thus proposes a skeptical approach to ethics: the subjects themselves must arbitrarily establish values, principles and norms, as they are not verifiable in an empirical or objective way.

3 From the Principles of Biomedical Ethics to Clinical Application

The principlism of Beauchamp and Childress, from the perspective of the analysis carried out by the National Commission, was later complemented by Albert Jonsen and Stephen Toulmin, who pointed out that, given that the *Belmont Report* had focused on the discussion of specific cases based on the application of the principles, it had proceeded in an unsatisfactory manner, since the use of the principles was not necessarily oriented towards an adequate application. They therefore propose a completely opposite approach, since the work of the Commission showed that it was possible to reach agreement on particular cases despite the existing differences in the ordering of the principles [23]. These authors propose a different methodological and procedural approach, along the lines of casuism. In this way, their proposal differs from the elaboration of an ethics with a claim to universal validity based on principles [25]. Thus, Jonsen and Toulmin updated the motivations of Aristotelian deliberation on individual cases, demonstrating that, in the application of ethical reasoning, the beginning of the procedure is not located in the principles, but in the concrete situations in relation to which the principles are applied. Thus, a type of judgments is obtained that, especially in the case of the application of experimental situations, moves in the sphere of probability, considering the principles as mere guiding maxims. Therefore, ethical decisions in biomedicine must be taken after reasonable and prudent deliberation.

From the above, and the experience he acquired in the work of the National Commission, Jonsen arrived at the formulation of a general procedural criterion that reversed the deductive sequence, establishing that the cases that biomedical ethics addresses should begin from the practical experience of application, and not from the search for general principles or normative compendiums that should then be applied deductively; the method that biomedical ethics follows should be projected in a context of clinical analysis [24]. This casuistic method is fundamentally based on making decisions given in a practical context, without seeking an argumentation with greater content, but rather one that provides prudential analysis related to decision-making. Therefore, the main orientation of greater weight in the relationship between doctor and patient will always be the one that highlights the autonomy of the latter, which in cases where it wants to prevail over the medical indications of beneficence must face certain questions that must be answered in the light of the analysis of the facts. It can be perceived that this concrete way of dealing with the practical problems that arise in the doctor-patient relationship makes it possible to establish evaluative judgments that help to classify positively or negatively the situation under analysis. The objective of this process allows distinguishing between relevant considerations (which operate with a certain degree of relevance in the ethical deliberation), important (those that after a thorough analysis of arguments seek to obtain clarity around the orientation), and decisive (those that consider the relevant factors are chosen for their determination). Jonsen closes with a criterion with which he classifies between "permitted" and "obligatory" actions, the first being those that can be carried out when, after a due analysis, there are no decisive considerations to carry out an opposition. On the contrary, obligatory considerations are those that arise in cases in which there are decisive alternatives in favor of one of the options.

This process of application of the method of principles is similar to that assumed by the procedural legal system of the Anglo-Saxon "common law," which, before guiding the judgment, is inclined to observe what occurs in the field of custom, thus revealing a clear inductivist accent. This can be seen in the way the courts of justice have proceeded in the history of American biomedical ethics, e.g., Baby M., Karen Ann Quinlan, Elizabeth Boucher, and others. In those proceedings, the juries (groups analogous to the Health Care Ethics Committees in which the whole of civil society is "ideally" represented) meet and, in the name of the whole community, resolve the matter, acting in an intermediate capacity between the entire community, on the one hand, and the courts of justice, on the other.

The scope of this type of reasoning around the principles of biomedical ethics, as we shall see, will contrast with the logic that will be present in the application of the same principles in continental Europe and in its projection towards Latin America, with the deployment that is followed in the Anglo-Saxon sphere, which is where, as we have seen, this area of applied ethics was born.

4 The Projection of Biomedical Ethics in the Hispanic Context

Given the recipients of this volume, and taking into consideration that a comparative analysis of the different spheres in which biomedical ethics is projected should be reflected in a text that confronts only models of argumentation from different latitudes, I will limit the following analysis to the Hispanic sphere. This is relevant from the moment in which the reflection around the research activity, experimentation and application of criteria around the principles of biomedical ethics experiences a special shade of differentiation with the deployment that is followed in the Anglo-Saxon area, which is where this area of applied ethics was born, as we have appreciated.

The argumentative structuring of Hispanic biomedical ethics will follow a hegemonic model that we will call "hierarchization of principles." With this, the organization of those at the same level, along the lines proposed by Frankena, will be confronted. To explain the above, we will especially follow the argumentative proposal of Diego Gracia, the most relevant bioethicist of Ibero-America. Gracia makes explicit in his evolutionary development an argumentation with which he structures in a more objective way the system of analysis of biomedical ethics, taking charge, precisely, of both principles and values, in the line of Reich's definition [15–18]. From Gracia's proposal, in turn, we will see how diverse continental European authors dialogue and debate with his proposal, to give reasons both in favor and against his argumentative scheme.

Gracia carries out the development of his bioethical analysis and, specifically of the subclassification of biomedical ethics, based on the philosophical background provided by his teacher, Xavier Zubiri. An in-depth study of this influence and its impact on bioethics and biomedical ethics is referred to in our extensive volume entitled *La bioética de Diego Gracia* [12]. From the influence of Zubirian fundamental ontology, Gracia raises a proposal of leveling or hierarchization of the principles of biomedical ethics by which he tries to overcome the American pragmatist approaches that point out as unnecessary the search for greater consistency in the foundation of objective criteria for this discipline. Gracia proposes a careful study of the functioning of moral reason, which acts as the foundation of the formulations of American bioethics. Working from the assumption that the rise of proceduralism has resulted in the inability to reach "substantive principles," this Spanish author seeks a more solid basis for the discipline, and not mere possible "courses of action" ([19]: 9–50).

In this sense, he turns to the contribution of John Arras, who proposes two types of modalities of action in the line of casuism, the second being the one that will be placed above the mere deductive application. In this way, this author distinguishes between: (1) an axiomatic casuistry—the art of applying to concrete cases any type of moral principles at hand; and (2) casuistry based on maxims—the art of making probable judgments on concrete situations, instead of certain judgments on universal and abstract matters ([2]: 20.35), where the solution is not reached by the formulation of moral axioms established a priori by the ethical theory, but by the convergence in the argumentation of people who by their prudence can lead to an adequate resolution of moral problems. Model 1 is of a formal and geometrical type, while model 2 is of historical and ecological features. This model 2, which has to be viewed from a perspective of historical rationality, requires not a mere deductive application, but one of a circumstantial nature.

This comparison is essential to confront the resolution of conflicts within possible courses of action in contexts of biomedical ethical deliberation. In this sense, those who are inclined towards the application of axiomatic approaches try to resolve complex situations by appealing to necessary and immutable principles, which in the context of countless contemporary situations are totally obsolete. On the contrary, those who argue that the way forward is through historical argumentation point out that it is impossible that procedural problems can be solved by appealing to the mere use of logical and a priori criteria, but rather that historical and a posteriori criteria must be employed, where principles take the form of frameworks or "reference systems" that help to maintain the consistency of the discipline by setting the framework within which hypotheses can subsequently be deployed. When conflicts do not interfere with the level of procedures, it is sufficient to resort to them to solve the problem. But when it comes to situations of greater depth, it is convenient to broaden the spectrum of considerations in order to formulate a new theoretical proposal illuminated by the experience that humanity has been creating. And the conflict lies precisely in the fact that when one seeks to establish permanent and universal procedural criteria, which are not open to the moment a posteriori, it is almost certain that one will arrive at the establishment of empty or merely formal procedures, which in the end, will not solve the ethical conflict. Therefore, after the devaluation of rationalism, there would be no other alternative but to construct hypotheses and theories within the framework of what Toulmin [32] has called "explanatory ideals," understood as referents of rationality and understanding, whose task is to account for the facts in order to provide a historical or ecological explanation of them.

It is from this analysis that Gracia will point to the understanding of the method of biomedical ethics on a double plane of levels, following a comprehensive Zubirian background. Together with Habermas, Gracia refers to the levels of autonomy and "self-realization." He understands the first meaning, in Kantian terms, as the absolute self-legislative capacity that consists of a universal moment of morality that is given prior to any a posteriori characterization, and that is located in the "disciplinary level" accepted by all human beings, contained in the fundamental human right Dworkinian that provides the wide framework of the reference system: all human beings are equal and deserve equal consideration and respect. In this sense, it is worth bearing in mind a relevant fact when specifying concepts: for Gracia, the location of the principle of autonomy in the biomedical scheme opened by the National Commission corresponds to the level of self-realization (not of autonomy), and from there he places this principle in Level 2 of the system.

Below or as a support for the level of self-realization is the disciplinary level, which accounts for the basal substrate of rights in which all coincide, so it is understood as the minimum moral stage that in the case of biomedical ethics is going to be expressed through the principles of non-maleficence and justice. With the first of these it is made explicit that all human beings are equal, deserving of equal consideration and respect "to the order of biological life," and, in the case of justice, "to the order of social life." That is to say, "when people are discriminated against, both biologically and socially, an injustice is being committed and, therefore, it goes against the universalization that the disciplinary level demands and demands" ([16]: 135). And next to the disciplinary level there is another of "non-disciplinary" cut, which is where it would correspond to the principle of autonomy of Belmont's scheme. In consideration to the universality expressed in the previous principles, the presence of particularity is what makes explicit this plane of historical, customary, personal, and non-transferable cut, which is the one that Habermas has baptized as of self-realization, equivalent to the biomedical ethical principle of autonomy. Next to this, the principle of beneficence can be located, which also develops in the plane of self-realization, since it cannot be imposed to achieve its purpose.

On the basis of the aforementioned reflection, the distinction between minimum and maximum principles arises in the Hispanic context, from principles developed by Adela Cortina in the field of general ethical foundations. In Spain in the 1970s, the return to democracy created a scenario in which many issues began to be debated in an incipient pluralism. This raised a problem specifically for ethics, and over time, for biomedical ethics, since these were precisely the disciplines oriented to opening discussion in contexts that necessarily created areas of confrontation in the public sphere. This is how Cortina's proposal arises in *Ética mínima*, by which she presents the distinction between a "minimum of consensual laws," which by means of its juridical positivization fixed the rules of civic life. Next to that normative minimum there would rise an outline of maximums tending to the conquest of happiness, in which "the values in which it is worthwhile to commit one's life" would be defined ([8]: 158 ss.). This is how a true civic ethics could be carried out based on the agreement of the moral minimums necessary to solve a moral pluralism.

It was a proposal by which the common moral minimums allowed establishing a limit to the power of the State, which should prevail only with respect to the citizen and not to the human being as such. In this way, Cortina would speak of ethics of *justice* or *of minimums*, and of ethics of *happiness* or *of maximums* (or *self-realization* in Habermas), in tune with deontologist authors like Rawls, Apel, Habermas or Kohlberg ([9]: 65). In Habermas' terms, the principles of the non-disciplinary level are those of autonomy and beneficence, understanding the first not as the universal and self-legislator element that constitutes the disciplinary level, in Kantian line, but the "own and particular project of life that each one is constructed" and that connects with the project of Habermasian self-realization, reason why we will notice here, yes, a difference with the location that Cortina makes of this principle. Then, in the non-disciplinary level, autonomy is understood as "the right of every person to realize his own ideal of perfection and happiness," being responsible for his decisions, since only autonomously can each define one's particular ideal of happiness and perfection. This demands a willingness not to impose models on those who seek ways of realizing maximums from the minimums established dialogically. This principle starts from the Kantian perspective, but opens towards an Anglo-Saxon approach by which autonomy is used to define what is beneficial for each one. In this way it is placed on the same level as beneficence, the other principle of private traits.

Going deeper into this classification of levels is that Gracia will speak more clearly of a certain hierarchy that is possible to appreciate between them, identifying the first level (non-maleficence and justice) with the juridical distinction that classifies the so-called "duties of perfect obligation" according to the denomination used by the philosophy of law of Kantian origin, as opposed to the duties of imperfect obligation, which he sees reflected in the second ones (autonomy and beneficence). The perfect duties establish the negative mandate of prohibiting the realization of certain actions, given their absolutely bad character. A different situation occurs with imperfect duties, in relation to which it is possible to consider degrees and the possibility of applying exceptional solutions that allow the option of leaving the doors open to the processes of biomedical deliberation.

Under the scope of formal law that all human beings are essentially equal, deserving of equal consideration and respect, it is possible to establish a distinction in the obligatory nature of compliance with the principles. In the case of the binomial non-maleficence and justice, even against the will of those who wish to resist, criminal law ensures respect for the former (non-maleficence) by coercively protecting certain legal goods that it considers especially valuable. By safeguarding the principle of justice, civil law does the same, ensuring respect for acts that generate binding obligations in both the public and private spheres. At the same time, the principles of autonomy and beneficence are considered of "lower objective rank" compared to the first two. The obligation to respect them does not have the coercive force to require their compliance even against the will of individuals, establishing between them and those of non-maleficence and justice a distinction similar to that between the common or universal good and the particular good. In case of confrontation, it is necessary to put the principles of minimums, or non-maleficence and justice (Level 1), before the principles of maximums, or autonomy and beneficence (Level 2).

From a liberal perspective, the former should be considered duties of perfect obligation, as they are managed publicly by means of a legal rule that must be complied with equally by all the members of a particular legal system, generating in others the correlative rights demanded by compliance with the specific rule, thus incorporating the distinctive note of "transitivity" of the act, as other persons are affected by the possible injury to their rights. On the other hand, the principles of autonomy and beneficence, being considered of private or "intransitive" management, respond to the beliefs and the particular situation of each one of the persons that conform the civil society, without being demanded its compliance by the legal system, since there is no risk of affecting transitively to presumed implied in the decision making. In this way, the universal principles of non-maleficence and justice precede, in terms of obligatory nature, the particular principles of autonomy and beneficence, which, being linked to purely intransitive acts, are not subject to legal sanction, but only to the moral examination that is possible to apply in the framework of an ethics of maxims.

In order to specify this distinction, Gracia's analysis focuses specifically on the difference that must be indicated, as mentioned above, between non-maleficence and beneficence, since, despite being classified differently, it gives the impression that they are two principles that manifest opposite sides of the same coin. From another perspective, Gracia argues that the distinction between the two has to do with the relationship of generality and particularity that occurs in the same direction, so they should be placed at two different levels. While non-maleficence is definable with universal and common criteria within the sphere of perfect duties, beneficence is located in the sphere of particularity and concreteness, in the plane of imperfect duties. In this sense, it is possible to say that the beneficent act operates on the basis of acceptance expressed through "informed consent," while non-maleficence is a moral and juridical imperative.

To clarify this point by means of an example, we can express that performing a blood transfusion to a Jehovah's Witness is not an act of beneficence, since in this case it goes against the system of values that the subject concretely possesses in the private sphere, in circumstances in which beneficence depends precisely on the latter, possessing a subjective connotation that contrasts with what happens in relation to the first principles. Therefore, what legitimizes the realization of such an act -in a legal framework that criminalizes active euthanasia or assisted suicide-, against the will of the individual in this case, obeys the application of the principles of non-maleficence and especially of *justice*, which operate in a superior sphere to those of autonomy and beneficence, without these being the principles that come into play.

In this sense, the ethics of minimums has custody of the so-called negative or perfect duties, which aim at the fulfilment of what is just, being guaranteed by the State through the exercise of law. On the other hand, the ethics of maxims promotes the realization of virtuous actions, or imperfect duties, which go beyond care because they aim to safeguard the minimum acceptable, and which open towards the realization and projection of a concrete values system. In this area, the State ensures compliance with the minimum acceptable moral standards; however, it is the family (together with other intermediate groups), from the privileged position/duty of guardianship and training, that builds and defines the moral dimensions of the members of society at its most basic level. They are, for example, called to promote exceptionality in an accepted way through the empowerment of its members towards optimal levels of achievement.

5 The Debate Around the Principles of Biomedical Ethics

With the development of this hierarchical scheme of principles, we could achieve, in the context of continental European reflection, the configuration of a general framework in a line radically opposed to that formulated by Beauchamp and Childress in *Principles of Biomedical Ethics.* It is what we could call a semi-rigid systematization of the principles of biomedical ethics within the hierarchical framework. With regards to the first level, this does not raise major problems, although, in relation to the second, it does allow for the possibility of pointing out exceptions. For this reason, from the perspective of hierarchical structuring of the principles, we will start from the assumption of doing no harm (primum *non nocere*) with non-maleficence—placing it as the first of the principles. Likewise, justice, understood in a positive way is the second of the principles that obliges from the transitivity, so in its binding character it follows the basic principle of non-maleficence—placing itself in a sphere that surpasses the remaining principles. In this way, and in relation to the minimum principles, we obtain a confirmation that establishes that in order to give to each his own (the exercise of justice), it is first necessary to do no harm.

In turn, from the perspective of the two principles of maximum or intransitive application, in the first instance, it is possible to prioritize between them in the context of an a-paternalistic scheme, which is the one that has progressively opened up with the establishment of biomedical ethics, In this sense, since the period of Modernity it is feasible to affirm that autonomy is necessary for beneficence to be given, since an act lacks the attribute of beneficence if it opposes the will of the patient. When this happens, we are no longer facing the realization of a beneficent act, but the opposite. That is why autonomy should be understood, in the context of contemporary personal interrelations, as a principle that, with rare exception, should be higher in the order of priority than beneficence. The criterion that marks the exception in these cases is the one suggested by Beauchamp and McCullough, who in *Medical Ethics* point out a clear synthesis in this respect. This approach, it seems to me, gives a greater degree of relevance to beneficence in a "strong paternalistic" context, and it can be perfectly taken up as an exception to the norm. In this sense, Beauchamp and McCullough argue that paternalistic intrusions should justify an exception to autonomy only when the user of the system is affected by a disease or by high-risk injuries that lead to medically justified decisions ([5]: 113 ff.).

Looking at the hierarchical scheme of principles from a critical perspective, in the Hispanic context certain authors have offered objections to the lexicographic table of the two levels identified by Gracia. Thus, it is pointed out that the principle of autonomy should be placed in Level 1 of the system, either as an independent principle, or as a moment of the principle of non-maleficence. In this sense, Adela

Cortina, following a Kantian interpretation, indicates that autonomy must be considered from the anthropological point of view as a constitutive principle of the human being, which is part of an ethics of minimums and, therefore, must be located in Level 1 of Grace's ordering ([10]: 237).

Meanwhile, Pablo Simón Lorda argues that autonomy consists of a duty of perfect obligation: a moral rule that responds to the correlative interpersonal fact of mutual respect. He proposes that the problem is in declaring autonomy as a principle, and not as a new symmetrical approach that relates moral subjects through non-maleficence, beneficence, and justice. Therefore, and in tune with Adela Cortina, the principle of autonomy should be articulated as a minimum duty, or a perfect obligation that is part of the principle of non-maleficence ([30]: 346).

In turn, Manuel Atienza challenges the non-justification in the hierarchization; it seems to him that identifying the two principles that make up Level 1 (non-maleficence and justice) as those that necessitate the "independence of the opinion, and will of those involved" constitutes a fallacy of *petitio principii* (i.e., begging the question), in which the justification for placing them at a higher level than the principles of autonomy and beneficence is not clear. In this same sense, bearing in mind that the principles of Level 1 directly express the content of the general Dworkinian principle of equal consideration and respect, Atienza points out that it would not be understandable that the opinion and will of an individual should count less than that of another, thus implying the need to incorporate it in Level 1, although for a different reason than the reason expressed by Cortina. Likewise, he points out that the classification in the two levels leads to sustain two criteria that he does not consider valid. In this sense, the interpretation that seems to emerge from Gracia is that, from a moral point of view, causing harm is worse than not doing good—he considers that it is worse to kill than to let die; therefore, it would justify the location of the principle of non-maleficence over that of beneficence, classifying the first in reference to the common good, while the second in relation to the particular good. For Atienza, this distinction would not be sustainable since the obligation not to do evil to another, and not to do good against the will lack justification because with this it is previously accepted that the "good" that a person considers as such has to remain in the subjective sphere, while the "evil" follows an objective appreciation, independently of what the subjects indicates as evil for themselves ([3]: 75–99).

Gracia's proposal of procedural hierarchization, however, has made it possible to overcome theoretical extremisms that have reciprocally denied one or the other level, with disastrous consequences in the history of humanity. For example, the extremism of political totalitarianism has been responsible for systematically annulling Level 1, trying to achieve a utopian ideal of general happiness based on the imposition of the law. On the other hand, the attempt of the extreme liberal utopia would impose Level 2 at the cost of eliminating the previous one. Thus, what results is a call for coherence reflected in the balance achieved accepting both levels of principles that correlate in a complementary way.

Diego Gracia himself, as a leading author in the definition of methodological criteria for decision-making within a biomedical ethics, subsequently advanced along a line that goes beyond the use of the simple ordering of the principles and has left them as mere companions of the material analysis that since 2000 was developed from the "biomedical axiology.“ Its aim is to relocate the methodology in a context or frame of reference that is not reduced to the semi-rigid hierarchical ordering of the principles of biomedical ethics and the weighting of foreseeable consequences, but also creates explicit criteria that lead to a more detailed evaluation of the options which must be weighed in the analysis. In other words, any attempt to move towards a humanization in the management of biomedical decision criteria using the system of the mere hierarchy of principles seemed to solidify in extreme decisions [13].

With this aim, a deliberative methodology close to the origins of the Aristotelian proposal is proposed. Through the reinstatement of the latter, Gracia goes in search of a path of knowledge by which the people involved can exercise in a pacific way and without coercion the analysis and exchange of their perspectives. At the end of this process, the quality of its course will be verified when it is found that the final solution, in the majority of cases, is different from the positions that the interlocutors put forward at the beginning of the sharing of their appreciations. In this sense, biomedical ethics can solidify a new form of relationship that should occur between the physician, the user of the health system and the respective institution that provides the medical service, all of whom should base their relationship on collective deliberation rather than on a calculated imposition of principles.

From a perspective of complementary analysis that contributes to the decision-making process, the same is to be expected as a means of resolving problems, and issuing the respective consultative reports that emanate from the Health Care Ethics Committees. In these instances, adequate preparation aimed at educating attitudes that help move from the inclination to defend personal biases to a deliberative path is an essential requirement. From the natural tendency of human beings to seek the custody and imposition of the criteria on which they stand and from which decisions are projected, towards a new way of facing the problems at stake. Overcoming tends to an adequate formation that presupposes a relative personal predisposition in this sense. Therefore, knowledge gained through deliberative discussion leads those involved to a process of analysis in which personal perspectives must be submitted to a peaceful and critical discussion. These types of intellectual definitions require a necessary moderation that must lead the participants, at least during the course of joint deliberationsve act, to consider individual positions in order to be open to the ways in which others perceive the discussion, and so be ready to make decisions in some future scenario.

The relevance of this new way of understanding the decision-making process in medicine goes beyond health contexts, being a praxis that necessarily affects the most delicate instances of social life. It is extremely important, therefore, to be aware that it is increasingly common to notice, in its projection to civil society, the weight and influence of the decisions that are adopted within the hospital environment. currently, the context in which the above would take place is observed in the way professionals and users of the health system are connecting, completely influenced by the autonomy exercised by the latter in a projective framework that has not yet reached its final implications.

Following the Gratian line, the constitutive elements of the methodological analysis that focused the attention of the first years of biomedical principlism (principles, circumstances, and consequences) must give way to consider a form of deliberative treatment of the same, now associating the incorporation of "value" criteria as a central theme. With the incorporation of values, it will be possible to deal more adequately and flexibly with ethical problems in more complex contexts and, in Weber's words, of "axiological polytheisms," such as the current one, especially in times when it is important to manage with greater narrative freedom. Thus, values allow, on the one hand, a detailed way of understanding the minimum necessary at the disciplinary level and, on the other, to make explicit how it is possible to develop the aspirations of autonomy and personal self-realization at the non-disciplinary level, or, as previously mentioned, within the framework of an ethics of maximums. Therefore, Gracia and other authors will make explicit the methodological application of a new biomedical ethics that relies on deliberation as a fundamental tool both for the preparatory moment of the analysis of the "facts" to be weighed and the confrontation of the values with which the judicative analysis of the facts is to be returned to and at the end of the decision-making process in the definition of the "duties" [20].

The clinical context in which these formulations are carried out has to do with the transition from care focused on the treatment of acute patients who require immediate action, to that dedicated to an increasing number of users who have increased their life expectancy through different procedures. The processes of health care would change to assume social aspects linked to dependency, and a series of situations related to the maintenance of chronic events or temporary survival. The variety of individual and collective ethical problems that health centers began to face should be reflected in the type of procedure and management of situations that would address the multiple facets of the quality of human life. The implication of the treatment of values in these situations is of greater force as a result of the effective possibility of prolonging the illness in a context of stability, which would entail the need to manage the deliberative methodology by which these problems would be faced, and avoid the tendency to clear conflicts through dilemmatic elucidations: the proposal that opens with axiological distinctions allows facing controversial situations in ethics without the need to reduce the problem in a dichotomous way. We find ourselves, therefore, in a new stage of treatment of biomedical relations that reflects a more mature and participatory approach to the controversial analysis that animates the deliberative process.

6 Conclusion

Since its origins, the emergence and development of biomedical ethics has meant an effective projection of practical ethics in the broad context of civil ethics. This happened precisely at a time when, thanks to the advances in biotechnology, a large number of situations arose that demanded methodological approaches that would take charge of an adequate operationalization of the decision-making processes in controversial situations related to the clinical treatment of the beginning, maintenance and termination of human life. In this scenario, there is a progressive distinction in the interpretation of how specialists must approach and finally solve the diverse biomedical conflicts that increase in quantity and complexity every day, and which requires necessary justification of the procedures of clinical ethical decision.

As we have described, scenarios that follow various justifications of deliberation methods allow us to contrast different criteria application strategies that explicitly show how this branch of applied ethics, in the end, does nothing more than reflect procedural models sustained in a fundamental ethical interpretation. Even within the same large organizational systems previously discussed, these disquisitions are manifested, indicating nuances that lead to the conclusion that from the same applied landing of ethical analysis, the consistency and plausibility of the foundations can be reconsidered. That is why, among other reasons, as Toulmin said, bioethics has enabled a reawakening of the moral discussion in the contemporary public sphere, positioning this practical branch of philosophy in a leading moment. It is not by chance that we are, as many mention, in the century of ethics; in this sense, biomedical ethics has a special relevance, advancing in the dialogic positioning of principles and values that today serve not only this area of applied ethics but also several of those we mentioned at the beginning of this section.

Debate within this area of ethics, which has provided the classic framework of principles and subsequently of values used by different applied sub-currents, is fully alive and in dialogue with the societies that constitute the changing moment that we are opening as a culture. To ethically optimize the performance of members, users, and health services is one of the great challenges of this relentlessly advancing century, and constitutes an open and challenging scenario for biomedical ethics.

References

1. Arnold DG (ed) (2009) Ethics and the business biomedicine. Cambridge University Press, Cambridge
2. Arras J (1990) Common law morality. Hastings Center Report, New York
3. Atienza M (1998) Juridify bioethics. Isonomía 8:75–99
4. Beauchamp T, Childress J (1979) Principles of biomedical ethics. Oxford University Press, New York
5. Beauchamp T, McCullough L (1987) Medical ethics. the moral responsibilities of physicians. Labor, Barcelona

6. Bonete E (2010) Neuroética práctica. Descleé de Brouwer, Bilbao
7. Callahan D (1990) Current trends in biomedical ethics in the United States of America. In: Bioethics. Issues and perspectives. Pan American Health Organization, Washington, pp 166–170
8. Cortina A (1994) Ética mínima. Tecnos, Madrid
9. Cortina A (1995) Ética civil y religión. PPC, Madrid
10. Cortina A (2007) Ética aplicada y democracia radical. Tecnos, Madrid
11. Dewey J (2008) Valuation theory. Siruela, Madrid
12. Faúndez-Allier JP (2013) La bioética de Diego Gracia. Triacastela, Madrid
13. Faúndez-Allier JP (2016) Diego Gracia's bioethics: from hierarchical principlism to axiological deliberation. In: Ferrer J et al (eds) Bioethics, the pluralism of the foundation. Universidad Pontificia de Comillas, Madrid, pp. 145–174
14. Frankena W (1963) Ethics. Prentice-Hall, Englewood Cliffs, New Jersey
15. Gracia D (1989) Fundamentos de bioética. Eudema, Madrid [2nd edition (2007), Madrid: Triacastela]
16. Gracia D (1991) Introducción a la bioética. El Búho, Bogotá
17. Gracia D (1991) Procedimientos de decision en ética clínica. Eudema, Madrid [2nd edition (2007), Madrid: Triacastela]
18. Gracia D (2004) Como arqueros al blanco. Estudios de bioética. Triacastela, Madrid
19. Gracia D (2007) Origen, fundamentos y étodo de la bioética. In: Martínez JA (eds) La bioética en la educación secundaria. Publications of the Ministry of Education and Science, Madrid, pp 9–50
20. Gracia D (2013) Construyendo valores. Triacastela, Madrid
21. Habermas J (1985) Moral conscience and communicative action. Peninsula, Barcelona
22. Jahr F (1927) Bio-ethics: a look at the ethical relations of man to animals and plants. Kosmos 21:2–4
23. Jonsen A, Toulmin S (1988) The abuse of casuistry. a history of moral reasoning. University of California Press, Los Angeles
24. Jonsen A, Siegler M, Winslade J (1998) Clinical ethics: a practical approach to ethical decisions in clinical medicine. Macmillan Publishing Company, New York
25. Pellegrino E (1999) The metamorphosis of medical ethics. a retrospective look at the last 30 years. In: Couceiro A (ed) Bioética para clínicos. Triacastela, Madrid
26. Potter R (1970) Bioethics: the science of survival. Perspect Biol Med 14:127–153
27. Potter R (1971) Bioethics: bridge to the future. Prentice-Hall, Engelwood Cliffs, New Jersey
28. Reich WT (ed) (1978) Encyclopedia of bioethics. McMillan and Free Press, New York
29. Ross D (1994) The right and the good. Sígueme, Salamanca
30. Simon P (1999) On the possible non-existence of the principle of autonomy. In: Sarabia J (eds) Bioethics, meeting place. Association of Fundamental and Clinical Bioethics/Zeneca-Farma, Madrid, pp. 343–350
31. The Belmont Report (1990) In: Clinical trials in Spain (1982–1988). Ministerio de Sanidad y Consumo, Madrid (Appendix 4)
32. Toulmin S (1977) Human understanding: the collective use and evolution of concepts. Alianza, Madrid

Development Ethics

2

Emilio Martinez-Navarro

Abstract

In an irreversibly globalized world, Development Ethics points out the main arguments, concepts and principles that can guide development policies within countries and also in international relations. Since it was born in the 1960s, Development Ethics has largely driven the transition from an initially very poor development concept to the present complex concept of sustainable human development.

Keywords

Development ethics • Globalization • Human rights • Human development • Sustainable development • International aid

1 The Birth of Development Ethics[1]

In the 1960s, a book on the *ethics of development* was published for the first time [30]. In those years, the expressions "developed country" and "underdeveloped country" became extremely popular in the midst of the commotion of the decolonization processes that were taking place on the African continent and elsewhere.

[1] An earlier version of this text was published under the title "Ethics for the development of peoples: a politics of dignified life without exclusions" in Mealla ([49]: 23–47)

This study is part of the Scientific Research and Technological Development Project FFI2016-76753-C2-1-P, funded by the Ministry of Economy and Competitiveness.

E. Martinez-Navarro (✉)
Universidad de Murcia, Murcia, Spain
e-mail: emimarti@um.es

A. Cortina and M. Correa Casanova (eds.), *Applied Ethics from Medicine to Humor*, Integrated Science 21, https://doi.org/10.1007/978-3-031-59009-2_2

The concept of "country development"[2] had emerged a few years earlier, associated with the good wishes expressed by many political leaders in the early years after the Second World War: the good wishes that the horror of such a war situation would never be repeated. These good wishes appeared to be linked to the idea that, if the various countries of the world truly achieved "development," wars would disappear and the Kantian dream of perpetual peace would finally come true [8, 13].

But those good wishes were soon affected by the so-called "cold war" between the Soviet communist bloc and the Western liberal bloc. In this new context, the idea of the development of peoples was used as one more element in the rivalry between the two great political blocs. In this way, the very notion of the development of countries was perverted and manipulated: because the primary ethical purpose of development processes—which, in principle, is to overcome poverty and promote a decent life for all inhabitants—was replaced by the propagandistic goal of "getting out of underdevelopment" by aligning oneself with the right side. Thus, in the 1950s and 1960s, multiple initiatives were carried out on both sides aimed at promoting rapid industrialization and "modernization" in many poor, recently decolonized countries, always under the slogan of rapidly achieving development and overcoming "underdevelopment." As Kenyan author Firoze Mangi has pointed out:

> [T]he new controllers of the state machinery assumed the role of "unique developers" and "unique unifiers" of society. The state adopted an interventionist role in the process of "modernization" and a centralizing and controlling role in the political sphere. Despite being born out of the struggle for legitimate pluralism and against the hegemony of the colonial state, social pluralism began to be frowned upon. The people's associations that had propelled the nationalist leadership to power gradually came to be seen as obstacles to the new god of "development." It was maintained that popular participation in decision-making was no longer necessary to determine the future. The new governments would be in charge of bringing development to the individuals. ([39]: 15)

Faced with such manipulation of the concept of development by various governments and groups defending their particular political and economic interests —in the context of the Cold War—critical voices emerged, denouncing these abuses as a deadly trap for the countries that were supposedly being "helped to emerge from underdevelopment." In fact, the French Dominican economist Lebret [36], the aforementioned Goulet [30], and Pope Paul VI with the Encyclical Letter *Populorum Progressio* [52] made pioneering contributions in the line of an ethics of development committed to pointing out that this new branch of ethics:

[2] The term "underdevelopment" was popularized from the speech of inauguration of the American president Harry Truman in 1949, which in the fourth point included the phrase: "We must devote ourselves to a new and determined program to achieve that the benefits of our scientific advance and of our world progress also serve for the advance and growth of the *underdeveloped* world," quoted in Sanpedro and Berzosa ([56]: 18).

> [W]ill have as its main pedagogical task to teach greater realism in relations between nations, between classes and between regions. In the matter of solidarity, human beings are still in their first phase, immobilized by their inability to overcome their individual and collective egoisms in order to achieve an unrestricted solidarity, which gives the principle of human fraternity its full dimensions, both in action and in being, not to mention official positions. If men and nations persist in their rejection of lived solidarity, they are preparing themselves for conditions in which solidarity will manifest itself as a common perdition. ([30]: 100)

Since those initial moments, a large number of international scholars from various branches of knowledge (anthropologists, sociologists, economists, engineers, political scientists, philosophers, etc.) have made interesting contributions in this field. The International Development Ethics Association holds regular meetings and symposia,[3] and publishes a specialized periodical: *Journal of Global Ethics*.[4] A growing number of specialists in development ethics have been developing and expanding a set of arguments that can be useful to anyone who intends to reflect rigorously on the assumptions and ethical implications of alternative models of development of countries. In what follows, we will give an account of some of the main components that articulate this field of ethical knowledge, and the most relevant issues of debate and research in this field. As ethical knowledge, it functions as a compass. By itself it has no power to change the world, but this *applied ethics* makes us much more attuned to what we should do in order to change it.

Development Ethics (DE) is a knowledge that guides the action of people and institutions involved in the tasks of promoting human development in harmony with the environment. As academic knowledge, it has two main components: (1) the contributions of various social sciences (sociology, political science, economics, etc.); and (2) the normative or evaluative reflection offered by moral philosophy or philosophical ethics. The most remote precedents of DE can be found in the debates and documents that gave rise to the 1948 Universal Declaration of Human Rights, and in the Declaration itself. In it we find the key elements of what we still understand today as *authentic development*, as opposed to self-interested falsifications. We can define development as *that social situation in which every person is guaranteed his or her basic rights and has real possibilities to exercise his or her obligations, and to carry out good life projects in harmony with his or her community and with the environment*. From this point of view, the goal of national development can be defined as the situation enjoyed by a political community in which it can be said that all human rights, and their corresponding obligations, are respected by all members of the community, to the extent that all have real capacity to promote permissible and ecologically sustainable life projects, without anyone having reasonably justified cause to feel excluded. This goal may seem utopian at first blush, but let us bear in mind that: (1) in recent decades there are some areas of the world that have managed to achieve it; and

---

[3] International Development Ethics Association (IDEA): http://www.development-ethics.org.

[4] *Journal of Global Ethics:* https://www.tandfonline.com/toc/rjge20/current.

(2) in those places where this scenario is still a dream, there are ethical reasons to denounce the abuses, and bring about the necessary changes to reach the desirable situation of authentic development, which is equivalent, in essence, to a situation of full compliance with Human Rights.

In fact, the UN Declaration of Human Rights appears in the same historical context in which the very notion of development emerged, dividing the world into two classes: that of the supposedly developed countries and that of the supposedly underdeveloped. In the UN's own Founding Charter of 1945, it is solemnly declared that one of the four purposes of the United Nations is to:

> ... achieve international co-operation in solving international problems of an economic, social, cultural or humanitarian character, and in promoting and encouraging respect for human rights and for fundamental freedoms for all without distinction as to race, sex, language or religion. ([65]: art. 1.3)

2 An Interdisciplinary, Pluralistic Academic Space Committed to the Cause of Humanity

The issue of the development of countries is enormously complex ([50]: 143) and concerns very diverse groups involved, starting with the very people who are supposed to benefit from the achievements of development. Consequently, *ethical* theorizing about development processes should not disregard what its protagonists can contribute: the beneficiary population, the planners of development policies, aid donor institutions, specialists from different academic disciplines, etc. If the point of view of any of the affected groups is completely disregarded, development will not be ethically adequate, but will result in some kind of "misdevelopment" that will sooner or later prove to be harmful to people, whether those of the present or future generations.

Interesting in this regard is the case of Denis Goulet, who in his pioneering monograph on the *Ethics of Development* (1965) includes the experience of vital commitment to the victims of development (which he sees as a process that often sacrifices individuals, institutions, and entire peoples) and of underdevelopment (perceived as synonymous with the precarious situation that arises from impoverishment, and ends with the premature death of people). From this point of view, Goulet affirms that the goal of development is that people can enjoy a *dignified* life:

> There is a certain agreement on the primary goods needed by underdeveloped humanity. These include, in the first place, the values of survival - the vital minimum of food, an adequate shelter to protect against the weather, the indispensable help for health, etc. In addition, in the various means of development, a certain convergence can also be discerned with respect to the needs of dignity, those that man must enjoy in order to lead a dignified life. ([30]: 19)

Indeed, from its first steps, DE has pointed out some *ends* that hardly require justification, such as those contained in Goulet's quotation. However, it has also provided a plurality of views on the *means* that can be used to achieve these ends. Does this pluralism of views on means indicate that DE fails to ethically guide action? Not necessarily. We must fully understand and assume that interdisciplinarity and worldview pluralism are valid and necessary elements in the context of the complexity that concerns all that is human; both are compatible with the honest search for a common truth that appears as the horizon of an "analogical community of thought" ([30]: 19) that unites people in the shared struggle to achieve the goals of the development of persons and societies. It is this shared commitment to a humanitarian cause that makes it possible to go beyond partisan and sectarian group interests.

From my point of view, the convergence of DE with other applied ethics, at least in some issues, is both inevitable and desirable. Because the analysis of the problems that affect the underdevelopment of countries leads to questions of *Ethical Economics* [6–9], of *Global Ethics* [18, 17, 42], of *Ecological Ethics* [15, 26, 61], of *Consumer Ethics* [12], of *Business Ethics* [12, 21, 24], of *Bioethics*—in the sense of "Global Bioethics" [1, 54]—and of *Legal Theory* [2, 34, 40]. In this sense, what finally counts is that the problems that prevent all the countries of the world from reaching true development are addressed and solved in the best possible way, and the dispersion of ethical knowledge does not have to impede this objective, at least if they do not become watertight compartments, but remain open to reality and respond creatively to the challenges it poses. For this, perhaps the notions of *citizenship* and *sustainable human development* are the keystones that allows the articulation of the ethical knowledge we have mentioned [11, 43]. In this sense, Amartya Sen has pointed out that:

> Citizenship is relevant for a number of different reasons. It can help individuals to behave more responsibly. It can provide reasons for environmentally friendly and generally more ethical behavior. But, going much further, the idea of citizenship brings to light the need to consider people as rational agents, not merely as beings whose needs have to be met or whose living standards have to be preserved. Moreover, it identifies the importance of public participation, not simply because of its social effectiveness, but also because of the value of that process itself. ([59]: 40)

3 An Increasingly Comprehensive Development Concept

From the beginning, DE has also been focused on a concept of development that does not confuse it with mere economic growth and the modernization of a society [19, 20, 48]. In this sense, David Crocker points out that among the sources of this Ethics we must count the reflections of "Gandhi in India, Raúl Prébish in Latin America, and Franz Fanon in Africa, who criticized colonial and orthodox economic development" ([16]: 76). The contributions of economists like Gunnar

Myrdal and Benjamin Higgins also contributed to show that the concept of development in those years was manifestly unilateral, insufficient and erroneous, when confused with the obsession for a fast economic growth, unequitable, and at any human and cultural cost. In contrast to this concept of so-called development, Goulet's Development Ethics maintained that:

> [D]evelopment is a total concept aiming at progress towards a human economy characterized by the progression of all men in all its dimensions. It adheres to the notion of development elaborated by the experts of the group "Economie et Humanisme" and is defined as follows: development is the discipline (of knowledge and action at the same time) of the passage, for a given population and for the sub-populations that constitute it, from a less human phase to a more human phase, at the fastest possible pace, taking into account the solidarity of sub-populations with populations. ([30]: 13)

The Centre Economie et Humanisme was founded in France in 1941 by Louis Joseph Lebret and, from the beginning of its multiple academic and cultural activities, it established itself as an international source of initiatives aimed at reestablishing an satisfactory relationship between ethics and economics, including the questions of development ethics. Many of the works carried out by this group of Catholic intellectuals had a decisive influence on the contents of Paul VI's encyclical [52] known as *Populorum Progressio*, the full title of which is *On the Necessity of Promoting the Development of Peoples*. This document already pointed to a large extent to what has subsequently been called *human development*. The fundamental principles of this document are expressed in the following way:

> To be free from misery, to find a more secure livelihood, health, a stable occupation; to share still more in responsibilities, free from all oppression and sheltered from situations that offend their dignity as men; to be more educated; in a word, to do, to know and to have more in order to be more; such is the aspiration of the men of today. And yet a great number of them are condemned to live in conditions which make this legitimate desire illusory. On the other hand, peoples who have recently attained national independence feel the need to add to this political freedom an autonomous and dignified growth, social as well as economic, in order to assure their citizens full human development and to take their rightful place in the concert of nations. ([52]: § 6)

The next step in the ethical deepening of the concept of development came with two polemics that arose in the 1970s: one around the responsibilities of the First World for the hunger and misery suffered by millions of people in the underdeveloped world; and the other around the priorities that economic policies should adopt to effectively help these people overcome misery.

In the context of the first polemic, the contribution of Peter Singer stood out. In his *Practical Ethics* (1979) he bitterly criticized a certain school of thought, headed by Garrett Hardin, who proposed letting the world's poor die of hunger and curable diseases as the only way for humanity to survive. According to this cynical pretension, if those of us who are free of misery were to help the impoverished, the result would be generalized impoverishment and the death of all. This "argument" to disregard the fate of almost a third of humanity has not found new, explicit support in the academic world; however, the fact is that rich countries' non-compliance with the Development Cooperation commitments actually reveals a double moral discourse on their part. On the one hand, their leaders seem to have accepted the recommendation approved by the UN General Assembly in 1974 on the need to donate at least 0.7% of their wealth to overcome those situations of extreme poverty. Still, several decades later, the average percentage of real aid is not even reaching 0.3% of the extraordinary amount of wealth hoarded by those same rich countries.[5]

From the point of view of DE, we can consider the contributions of Singer [60], Pogge [53], Schweickart [57], and García Delgado and Molina [25] regarding the ethical responsibilities of rich countries and rich people with respect to the situation of extreme poverty of millions of people, as valuable elements of criticism of the actual prevailing concept of development. These authors have placed great emphasis on the urgent need to put an end to the hypocrisy and bad practices of governments and world organizations—especially the WTO—in their relations with impoverished countries. Examples of such malpractices include: (a) recognizing coup-plotting and corrupt governments as legitimized to sell the natural resources of the country they militarily control, and to burden it with loans that generally do not benefit the population at all, (b) sheltering such rulers in their demands to have bank accounts in "tax havens," which are scandalously tolerated by the international community; (c) tolerating and even encouraging ample opportunities for labor exploitation, environmental pollution, and multiple violations of human rights in the name of "market freedom," in many poor countries, while at the same time maintaining protectionist policies towards products from other poor countries; (d) not donating to poor countries a minimal part of their wealth that could be enough to end extreme poverty ([53]: 21).

For their part, García Delgado and Molina [25] rightly argue that: (1) it is urgent to change the asymmetries in existing transactions between developed and developing countries, not only financially but also in terms of trade, environment, and technology, (2) it is necessary to abandon the neoliberal paradigm of the "minimal state," whose consequences in terms of sustainable human development have been disastrous, and replace it with a new paradigm of "responsible state," which ensures the redistribution of income, the preservation of public goods, the promotion of competitiveness with social cohesion, and generates decent employment for all.

---

[5] This approach, known as the "lifeboat ethic," was illustrated with this metaphor in Garret Hardin's article "Living in a Lifeboat," first published in the October 1974 issue of *Bioscience* magazine: *if we let the poor get into our boat, we will all sink.* Hardin's dilemma is false. The reality is not that the rich are in a crowded boat where there is no room for more people, but rather that we are sailing on a luxury cruise ship on which there are also poor people who survive by working hard for the rich, and on which there would be room for all the castaways swimming around if only the richest would agree to share a part of their wealth.

With regard to the priorities that economic policies should adopt with a view to the development of peoples, in the 1970s the contributions of Galtung [23], Max-Neef [47], Streeten [63], and many others rightly point out the imperious need to put the satisfaction of basic human needs as the main objective guiding development policies and development aid. These kinds of contributions paved the way for a new concept of development that appeared in the early 1990s: *human development*. But before going into detail on this notion, we must account for the novelty introduced in development issues since the late 1980s: *sustainable development*.

Indeed, since the publication in 1987 of the report prepared by the World Commission on Environment and Development, chaired by Dr. Brundtland, the most important ethical controversy has focused on the possibilities and imperatives posed by the worrying ecological situation affecting the entire planet. The report clearly showed that the development model prevailing up to that time was unsustainable. Sustainable development was defined as *development that meets the needs of the present generation without compromising the ability of future generations to meet their own needs*. At the end of the document, the Commission states that:

> [T]he issues we have raised in this report are inevitably of far-reaching importance for the quality of life on Earth - indeed, for life itself. We have sought to show how human survival and well-being may depend on the success of efforts to make sustainable development a global ethic. [68]

From the point of view of DE, questions related to care for the environment came to the forefront of study and debate with contributions from a large number of theoreticians who have insisted on the possibility of making environmental sustainability—in its dual aspects of preservation and recovery from damage to the environment—compatible with *human development* without exclusions [32, 41, 55]. The central argument is that one will not even be possible without the other. This will be more easily understood by examining the ethical implications of the notion of human development below.

The concept of country development as *human development* has been defined as the process that leads to the expansion of people's choices and capabilities, which takes the form of an improvement in: (1) life expectancy (the main components of which are health and security from violence); (2) education (measured in terms of literacy and schooling); and, (3) access to the resources necessary for a decent standard of living (measured in per capita income at purchasing power parity). This approach to development has been strongly promoted by the authors of the *Human Development Report*, published annually by the United Nations Development Programme (UNDP) since 1990. There is a fairly widespread consensus on the importance of the contribution of the Indian economist Amartya Sen in the paradigm shift implied by the emergence of the concept of *human development* [33]. Sen calls his proposal *development as freedom*, after having worked for many years on the question of the relationship between ethics and economics.

Sen's main idea is that development is not so much a matter of increasing income or wealth, rather a progressive expansion of human capabilities that allows people to enjoy sufficient freedom to carry out those life projects they value [58]. Let us summarize some of Sen's theses that have laid the theoretical foundations of the concept of human development:[6]

- The achievements of the last two centuries in terms of economic development and the spread of liberal democracy, the discourse on human rights, and the achievement of greater life expectancy at birth are generally positive; at the same time, the misery, oppression, and environmental degradation in impoverished countries are deplorable and outrageous.
- Overcoming these scourges requires, among other things, a new approach to development which aims, as a priority, to expand the real freedoms of all people.
- The active participation—with *agency*—is needed of each and every person who aspires to development.
- Development is the result of an integrated process in which economic, social, and political activities are deeply interconnected.
- The expansion of freedoms depends on a multitude of factors, such as education, health, political rights, and access to public debate, all of which are relatively independent of economic growth rates.
- Development is opposed to: famines, the current malnutrition of millions of people, the lack of access to health care, clean water and sanitation, the lack of basic education, the lack of employment or any economic security, the lack of equal treatment for women that would allow to enjoy the freedoms enjoyed by men, the lack of social and economic conditions to achieve a certain longevity that is viable today, and the lack of democratic freedoms.[7] To this point of view by Sen we could add that the goal is that everyone can enjoy the freedoms enjoyed by men. Not only that: men should also be freed from the shackles of gender roles. The rights of the LGTBIQ + collective should also be included in the application of development ethics, in order to follow principles such as *leave no one behind*.
- The *development as freedom* approach does not imply a single model of development for all parts of the world or all possible scenarios, since freedoms are diverse, and the ways of ordering them in a coherent system are many.
- Contrary to what has often been said, measures to promote *human development* are not a luxury that can only be afforded by countries that are already rich.

---

[6] Sen's contribution to development disciplines, including DE, is widely recognized and has given rise to new academic and civic initiatives that use the *capability approach* as a central reference, such as the Human Development and Capability Association.

[7] There is an abundant literature on the need to build on the advances made in gender studies in village development work. See López and Alcaide [38].

Most of the countries that have become rich began by giving priority to education, health and, in general, attention to people's basic capabilities; this has been a decisive factor in the dynamics of economic development itself.

- The democratic system, in a broad sense, constitutes an essential element of the development process for three reasons: (1) its direct importance as an expression of basic capabilities, such as political and social participation; (2) its instrumental role in improving the possibilities of individuals to express and defend their demands for political attention (including their economic demands); and, (3) its constructive role in the conceptualization of needs (since needs are not the same in all latitudes and in all cultural groups, but are always specific to each geographical and social context). To this must be added the permissive role of political freedoms, which allow and encourage public discussion. This discussion can prevent many economic disasters, although its concrete usefulness will depend to a large extent on the priorities of citizens, and the use they make of the political means at their disposal.
- In development processes, a focus on women's active *agency* is highly relevant, both for women's own well-being and for its impact on other areas, such as child survival and fertility reduction. Recognition of women's leadership is perhaps the most urgent requirement in the political economy of development, and must be a priority in DE.
- As for population growth, although Sen believes that the negative impact of population growth is often exaggerated, he also believes that fertility rates in most developing countries need to be reduced anyway. But the way that has been shown to be most appropriate for this is not coercive imposition and manipulation, but measures that increase women's freedom (through education, employment, health, etc.), and the responsibility of families in voluntary planning.
- Another controversial issue is that of the supposed opposition between the values that sustain freedoms and the indigenous values of Asia, Africa, and other regions of the world ([58]: chap. 10). In Sen's view, there has never been a homogeneity of indigenous values in either the East or the West, and just as there have been authoritarian positions in the history of the West, there have also been non-authoritarian positions in the history of the East. In development processes, no attention should be paid to unfounded generalizations that seek to associate authoritarianism with "true local values" and thereby elevate unscrupulous leaders who adopt a repressive attitude towards dissidents within their own local culture.
- It has also been said that it is not possible to rationally program a process of social development. Different reasons have been put forward, but Sen rightly refutes them, and insists that there can be no development without an overall concern for the process of increasing individual freedoms, accompanied by a social commitment to help bring it about.
- There are two major models of the process leading to the development of countries: BLAST (acronym for *blood, sweat, and tears*, a development model that requires "blood, sweat and tears") and GALA (acronym for *getting by, with a*

little assistance, an alternative development model to the previous one, under the motto "going together, with a little help"). The latter is clearly preferable, both from an ethical point of view and from a purely economic point of view, because supporters of the BLAST model tend to forget that mutual support and cooperation provide environments of trust and personal advancement that are not only desirable in themselves, but also generate greater productivity in the medium- and long-term.

4 Remaining Challenges for Development Ethics

The discourse surrounding the development of countries has been controversial ever since the term *development* began to be widely used in the mid-twentieth century—at least as controversial as the concept of *globalization* is *today*—because there are different ways of understanding development, both as a *goal* to be achieved and as a *process* by which progress is made towards the goal. In both aspects, there is a need for ethical reflection that can guide reforms, medium- and long-term policies, and the decisions to be taken by the agents who have responsibilities in the tasks of development.

A summary of progress in the two areas mentioned above may be useful in order to have a clearer idea of the road travelled so far, and the possibilities for progress in the immediate future.

In terms of development as a *goal*, the initial concept of development as mere economic/GDP growth, and the application of the macroeconomic measures of the so-called "Washington Consensus" [62] has been replaced by a more ethical vision of development in terms of *sustainable human development*. The latter is a broad and ambitious goal that can be broken down into several partial objectives, addressing many other fundamental dimensions of human life, such as food, health, education, housing, employment, citizen security, the exercise of democratic rights, and the protection of non-harmful cultural identity [41, 64, 66]. There is broad consensus among those of us working in DE that the *eradication of extreme poverty* is the main priority [44, 66, 67]. This is not at odds with the existence of some discrepancies regarding the definition of extreme poverty, the identification of its primary causes, and above all, the main ways of combating it. But it is no trivial matter that there is general agreement on the priority of fighting poverty without blaming the poor for it [14, 46].

In order to achieve the ambitious and feasible goals of development, it is necessary that all agents working in such tasks adopt certain ethical attitudes congruent with the sense of authentic development. Hence the interest shown by various DE authors who point out that it is the responsibility of professionals and institutions [27, 29, 31, 35] to take measures to prevent corruption, and to correct it when it appears.

In addition to the necessary change in personal and institutional attitudes required to advance towards the objectives of sustainable human development, there is also a broad consensus on the need for political changes and profound reforms of the international system [4, 25, 51]. This leads to the need to activate the conscience of citizens around the world through networks of humanitarian organizations that promote actions of vindication, denunciation, *education for development* [5], and cooperation through development projects [22, 45]. The fact that the United Nations proclaimed the *Right to Development* in 1986 poses the challenge for DE to make a serious and continuous contribution to the achievement of this right for all the countries of the planet: from the academic sphere, civic associations and social movements. This is so because the solemn declaration of the right to development is still a matter of good wishes that clashes with the harsh reality of global exclusion, abysmal inequality and chronic underdevelopment that afflicts more than a billion people, not to mention the serious unsustainability that afflicts the development model adopted by the "developed" countries (we should call them "underdeveloped," as I argued in [41]).

In this context, the vitality of DE is a beacon of hope in the midst of a dark storm of inequalities, waste of resources, and serious systemic risks that affect us all.

On the other hand, the development as a *process* leading to the supposedly desired goal of sustainable human development, both locally and for the planet as a whole, poses the greatest challenges to DE; in this area there are sharp disagreements about the role and size of the state, the size and regulation of markets, the responsibilities of citizen associations (the so-called *civil society*), etc. Philosophical diversity on these issues [3] is a cultural richness, but at the same time it carries with it a mortgage on the dispersion of forces, exploitable by the powers that be to maintain the *status quo*. The legitimate internal divisions between utilitarians, liberals, libertarians, neo-Marxists, and other philosophical traditions should not prevent the necessary convergence on issues that affect the present crisis, and the possibility of a shared future (without arbitrary exclusions) among human beings, and between them and other species of living beings.

However, over and above the inevitable ideological discrepancies, this convergence in the essentials is not at all easy; the ideological frames of reference do not allow us to reach a consensus in the diagnosis, nor in the treatment or mode of resolution of the problems posed by underdevelopment." In this sense, those of us who cultivate DE have the responsibility to: (1) reconcile opposing positions as far as possible; (2) denounce the hypocrisies and hidden interests that are often revealed in certain philosophical discourses, and in some supposedly scientific discourses; (3) encourage, with the best arguments, the implementation of urgent and lasting solutions to the problems of misery, exclusion, the mass extinction of animal and plant species, global warming, etc.

References

1. Alleyne G (2002) Health, ethics and development. In: Kliksberg B (ed) Ethics and development. The marginalized relationship. El Ateneo, Buenos Aires, pp 211–222
2. Angulo Sanchez N (2005) The human right to development in the face of market globalization: concept, content, objectives and subjects. IEPALA, Madrid
3. Arnsperger C, Van Parijs P (2002) Economic and social ethics. theories of the just society. Paidós, Barcelona
4. Black M (2003) What is international development? Intermón Oxfam, Barcelona
5. Celorio G, López de Munain A et al (2007) Dictionary of education for development. Hegoa, Bilbao. http://www.hegoa.ehu.es/es
6. Chang H-J, Grabel I (2004) Reclaiming development. Zed Books, London (forthcoming: Reclaiming development. Un manual de política económica alternativa. Barcelona: Intermón-Oxfam, 2006)
7. Chaves Ortíz JA (1999) From utopia to economic policy. For an ethics of economic policies. San Esteban /Edibesa, Salamanca/Madrid
8. Conill J (2007) Universal ethics of cooperation and development in solidarity. In: Jongitud J (ed) Ethics of development and social responsibility in the global context. Universidad Veracruzana, Veracruz, pp 27–44
9. Conill J (2004) Ethical economy horizons. Aristotle, Adam Smith and Amartya Sen. Tecnos, Madrid
10. Cortina A et al (1994) Company ethics. Trotta, Madrid
11. Cortina A (1997) Citizens of the world. Towards a theory of citizenship. Alianza, Madrid
12. Cortina A (2002) For an ethics of consumption. Consumer citizenship in a global world. Taurus, Madrid
13. Cortina A (2006) Development ethics: a path to peace. System 192:3–18
14. Cortina A (2017) Aporophobia: the rejection of the poor. A challenge to democracy. Paidós, Barcelona
15. Crabtree A (2018) In: Drydyk J, Keleher L (eds) Sustainability and climate change, pp 209–225
16. Crocker D (2003) Globalization and human development: ethical approaches. In: Conill J, Crocker D (eds) Republicanism and civic education beyond liberalism? Comares, Granada.
17. Dower N (1998) World ethics. The new agenda. Edinburgh University Press, Edinburgh
18. Dower N (2018) In: Drydyk J, Keleher L (eds) Global ethics, pp 17–28
19. Drydyk J, Keleher L (2018) Introduction. In: Drydyk J, Keleher L (eds) What is development ethics?, pp 1–13
20. Drydyk J, Keleher L (eds) (2018) Routledge handbook of development ethics. Routledge, London. Open Access: https://doi.org/10.4324/9781315626796
21. Enderle G (2002) Some links between corporate ethics and development studies. In: Kliksberg B (comp.) Ethics and development. The marginalized relationship. El Ateneo, Buenos Aires, pp 345–372
22. Etxeberria X (2004) Ethics of humanitarian aid. Desclée de Brouwer, Bilbao
23. Galtung, J (1978/9) The new international order and the basic needs approach. Alternatives 4:455–476
24. García Marzá D (2004) Business ethics. From dialogue to trust. Trotta, Madrid
25. García Delgado D, Molina MG (2006) Ethics and development. The conflict of interpretations. In: Scannone JC, García Delgado D (comps.) Ética, desarrollo y región. Towards an integral regionalism. Fundación Centro Integral Comunicación, Cultura y Sociedad CICCUS, Buenos Aires, pp 51–116
26. García Gómez-Heras JM et al (1997) Environmental ethics. Tecnos, Madrid
27. Gasper D (2004) The ethics of development: from economicism to human development. Edinburg University Press, Edinburg

28. Goulet D (1971) The cruel choice: a new concept in the theory of development. Atheneum, New York
29. Goulet D (2006) Development ethics at work: explorations 1960–2002. Routledge, London & New York
30. Goulet D (1965) Ética del desarrollo. Estela and Barcelona: IEPAL, Montevideo (there is Portuguese translation, *Etica do Desenvolvimento*. Sao Paulo: Livreria das Cidades, 1966)
31. Goulet D (1995) Development ethics. A guide to theory and practice. Zed Books, London (available in Spanish translation by Gabriel Rosón: *Ética del desarrollo. Guía teórica y práctica*, Madrid, IEPALA, 1999)
32. Goulet D (1995) Authentic development: is it sustainable? In: Trzyna T (ed) Defining and measuring sustainable development. International Center for the Environment and Public Policy & California Institute of Public Affairs for IUCN—the World Conservation Union, pp 44–59
33. Griffin K (2001) Human development: origin, evolution and impact. In: Ibarra P, Unceta K (eds) Ensayos sobre el desarrollo humano. Icaria, Barcelona, pp 25–40
34. Gómez Isa F (1998) The right to development: between justice and solidarity. University of Deusto, Bilbao
35. Kliksberg B (2004) More ethics, more development. Editorial Temas, Buenos Aires
36. Lebret L-J (1961) Concrete dynamics of development. Les Editions Ouvrières, Paris
37. López I, Alcalde AR et al (1999) Gender relations and development. Towards equity of cooperation. Instituto Universitario de Desarrollo y Cooperación—Los libros de la Catarata, Madrid
38. López I, Alcalde AR et al. (coords.) (1999) Relaciones de género y desarrollo. Hacia la equidad de la cooperación. Madrid: Instituto Universitario de Desarrollo y Cooperación – Los libros de la Catarata
39. Mangi F (2000) The depoliticization of poverty. In: Mangi F et al Desarrollo y derechos humanos. Intermón, Barcelona, pp 12–34
40. Marks (2018) In: Drydyk J, Keleher L (eds) The right to development. Ethical development as a human right, pp 266–280
41. Martínez Navarro E (2000) Ethics for the development of peoples. Trotta, Madrid
42. Martínez Navarro E (2003) Cosmopolitan ethics for international coexistence. Daimon J Philos 29:171–182
43. Martínez Navarro E (2007) Development as full citizenship. In: Jongitud J (ed) Ethics of development and social responsibility in the global context. Universidad Veracruzana, Veracruz, pp 139–169
44. Martínez Osés PJ (2005) Millennium development goals: can we end poverty? PPC, Madrid
45. Martínez Navarro E (2006) Denunciation and cooperation: the ethics of solidarity organizations for development. Ignacio Ellacuría Forum, Murcia. http://www.foroellacuria.org
46. Martínez Navarro E (2002) Aporophobia. In: Conill J (eds) Glossary for an intercultural society. Bancaja, Valencia, pp 17–23
47. Max-Neef M (1993) Human scale development: conception, application, and further reflections. Apex Press, London
48. Mealla E (2006) The return of development. In: Scannone JC, García Delgado D (comp.) Ética, desarrollo y región. Towards an integral regionalism. Fundación Centro Integral Comunicación, Cultura y Sociedad CICCUS, Buenos Aires, pp 117–141
49. Mealla E (comp.) (2013) Education and ethics for another development. La Crujía, Buenos Aires
50. Morín E (2002) Are we on a Titanic? In: Kliksberg B (comp.) Ethics and development. The marginalized relationship. El Ateneo, Buenos Aires
51. Mutsaku K (2003) Development and liberation: possible utopias for Africa and Latin America. ITESM, Mexico
52. Paul VI (1967) Encyclical Letter *Populorum Progressio*. http://www.vatican.va/holy_father/paul_vi/encyclicals/documents/hf_pvi_enc_26031967_populorum_sp.html
53. Pogge T (2002) World poverty and human rights. Polity Press/Blackwell, Oxford (translated by E. Weikert García: *Poverty in the world and human rights*. Barcelona: Paidós, 2005)
54. Puyol A, Rodriguez H (eds) (2007) Bioethics, justice and globalization. Erein, Donostia

55. Riechmann J (1995) Sustainable development: the struggle for interpretation. In: Riechmann J (eds) From economics to ecology. Trotta/Fundación 1° de Mayo, Madrid, pp 11–36
56. Sanpedro JL, Berzosa C (1996) Awareness of underdevelopment twenty-five years later. Taurus, Madrid
57. Schweickart D (2002) After capitalism. Rowman and Littlefield, Lanham
58. Sen A (1999) Development as freedom. Ramdom House, New York (Spanish translation by E. Rabasco and L. Toharia, *Desarrollo y libertad*. Barcelona: Planeta, 2000)
59. Sen A (2005) Reanalyzing the relationship between ethics and development. In: Kliksberg B (comp.) The pending ethical agenda of Latin America. Fondo de Cultura Económica, Buenos Aires, pp 35–41
60. Singer P (2000) One world. The ethics of globalization. Yale University Press, New Haven (Spanish translation by F. Herreros, *Un solo mundo. La ética de la globalización*. Barcelona: Paidós, 2003)
61. Sosa N (1990) Ecological ethics. Libertarias, Madrid
62. Stiglitz, J (2002) More instruments and broader goals for development. Towards the post-Washington consensus. In: Kliksberg B (ed) Ética y desarrollo. La relación marginada. El Ateneo, Buenos Aires, pp 45–99
63. Streeten P et al (1981) First things first: meeting basic needs in developing countries. Oxford University Press, London (English translation: *First Things First: Meeting Basic Needs in Developing Countries*. Madrid: Tecnos/World Bank, 1986)
64. UN (1986) Declaration on the right to development. http://www.unhchr.ch/spanish/html/menu3/b/74_sp.htm
65. UN (1945) Founding charter. http://www.un.org/spanish/aboutun/charter/index.htm
66. UN (2015) Sustainable development goals. http://www.un.org/sustainabledevelopment/es/objetivos-de-desarrollo-sostenible/
67. UN (2000) Millennium development goals. http://www.unmillenniumproject.org/reports/spanish.htm
68. World Commission on Environment and Development (1987) Our common future. Alianza, Madrid

Ecological Ethics: Critical Review and Proposed Rationale

3

Alvaro Ramis Olivos

Abstract

The text addresses the increasing porosity of the boundaries that define ecological ethics as a specific field, as well as its growing complexity due to its intersection with various disciplines. It examines several ethical currents, from traditional to contemporary, including ecofeminism, utilitarianism, and deep ecology, among others, highlighting both their contributions and limitations. A common ethical-discursive foundation for ecological ethics is proposed, aiming to overcome the aporias detected in the different analyzed currents.

Keywords

Ecological ethics • Planetary boundaries • Complexity • Intersection • Ethical currents • Utilitarianism

1 Introduction

The boundaries that could delimit ecological ethics as a specific field and totally differentiated from other specific ethics are increasingly porous, so it is worth making a critical review of its main currents and perspectives, trying to single out their contributions. The ethics applied to the ecological field constitute an academic discipline that is increasingly diverse and complex to analyze. In this diagnosis there is a greater specialization of their analytical perspectives due to a strong interdisciplinary confluence, which leads them to interact with the field of sciences, economics, politics, bioethics, or in general with the ethics of human institutions, which have assumed environmental dilemmas with increasing centrality.

A. Ramis Olivos (✉)
Universidad Academia de Humanismo Cristiano, Santiago, Chile
e-mail: aramis@academia.cl

A. Cortina and M. Correa Casanova (eds.), *Applied Ethics from Medicine to Humor*, Integrated Science 21, https://doi.org/10.1007/978-3-031-59009-2_3

This chapter will review some relevant authors and approaches, starting by outlining the specific rationale that should integrate ecological ethics. This common basis is rooted in the epistemology of ecology as a biological science. Next, the chapter will outline the characteristics of ecological ethics typical of traditional societies, where normative frameworks of moral character had adaptive functions to the environment, establishing through symbolic-religious traditions certain ecological functionalities increasingly valued by Western society.

The analysis continues, establishing a distinction with modern societies, where techno-scientific reason and contemporary polytheism values prevail. In this way, we will review Norton's proposal referred to a "weak anthropocentrism," Naess's "deep ecology," some approaches to ecofeminism, Singer's utilitarian extensionism, the biocentric ethics of authors such as Taylor and Regan, and finally the Latin American social ecology of Godynas and Acosta. In all these perspectives an explicit ethical positioning emerges, although it does not always result in the systematic formulation of an applied ecological ethics.

The analysis of these authors and approaches is formulated from a critical perspective, trying to highlight their novelties and contributions, but also showing their limitations and contradictions, either at the level of their abstract and argumentative formulation, or in the field of their applicability. The chapter concludes by proposing an ethical-discursive foundation common to ecological ethics, based on the proposal of Karl-Otto Apel and Adela Cortina. This proposal seeks to provide elements of complementation and overcoming of the aporias that are detected in the different currents analyzed, especially at the level of its foundation.

2 The Specific Rationality of Ecological Ethics

Ecology, understood as a branch of biology, is the scientific discipline that focuses on the study of ecosystems. It is a field of research initiated by the work of the Prussian naturalist and philosopher Ernst Haeckel, who was the first to use this concept in a proper sense in his work *General Morphology of Organisms*, published in 1866 ([55]: 138). Haeckel integrated the Greek words *oikos* (dwelling, house or home) and *logos* (study, treatise) giving rise to the expression "study of the house" or "of the home." His interest was to emphasize the systemic analysis of the relationships between biotic and abiotic organisms, such as matter and energy, and their mutual influences and transformations. This relational field forms the environment of biocenotic communities, where the diverse species that share a biotope or biological territory interact.

The Haeckelian image of the *oikos* refers to a sphere of complex links, under the metaphor of the common house of biological species, situated in an abiotic context that actively interacts with them. It is not about the "management of the house," as understood by economics, which since the Greek classics has studied the *oikos* from the productivist sphere. The ecological *oikos* is, above all, the relational sphere that is woven in biocenotic communities, and that explains the ties of dependence and influence between their individuals, in permanent interaction with climatic, physical and chemical factors, such as temperature, humidity, or light. In this way, the ecological approach introduced by Haeckel modified the study of species. While Darwin had focused on the analysis of their specific singularities, with individual emphasis, under the ecological paradigm, the analytical criterion came to focus on the networks that species establish among themselves and with their surrounding abiotic environment, since it is this context that determines their functional characteristics.

In this way, ecology—as a biological discipline—broadened awareness of the relationality of the natural environment, which not only establishes links between plants and rocks, animals and atmospheric gases, microorganisms and oceans, but also actively involves human beings. For this reason, he emphasized the importance of systematically studying the anthropic influence on ecological balances.

If ecology were understood exclusively as a descriptive dimension, there would be no need for a specific ecological ethics. It would be the same as proposing an anatomical ethics or a botanical ethics, since ecology, understood as a merely descriptive discipline, would not demand from practical reason anything more than an ethics of ecological research, similar to that demanded by any field of scientific research. However, ecology is an inherently normative science, since it seeks to explain and intervene in the course of biokenotic processes. This dimension stems from the intersubjective character of its analysis, since human beings are an inseparable part of the biomes under analysis. It is this aspect that underlies the need for an ecological ethic, understood as the process of critical accountability with respect to the effects of anthropic activity on biomes and ecosystems.

During the twentieth century there has been a progressive increase in the demand for a specifically ecological ethic. This has developed in a reactive way to the catastrophes and processes of environmental degradation that have accompanied the economic, technical, and productive growth of our era. This process, long described and characterized by various international forums such as the document "The Limits to Growth," written by MIT on behalf of the Club of Rome in 1972, the "Brundtland Report" (1987), the "Rio Declaration" (1992) and the "Fourth Report of the Intergovernmental Panel on Climate Change" (2007), has warned about the negative externalities of productive activities on the environment and has opened a debate on the sustainability of development models.

But it is worth noting that the ecological approach initiated by Haeckel has recognized that this reaction, born *after* destructive emergencies, has also been accompanied by a clearer awareness of the dimensions that link humanity with its environment in an integral way, which is why ecological ethics can be described as an important turn in the moral rationality of our time.

As Garrido Peña ([23]: 31) observes, the new ecological science introduced an epistemological rupture since it presupposed elements that anticipated contemporary systemic thinking. For the nineteenth-century encyclopedism, nature was above all an objectual space that had to be codified and catalogued by means of infinite taxonomies that distinguished, before linking, species and their processes. For ecological science, on the other hand, nature is conceived as a connected web of influences and links that do not depend on human gnoseological processes. The emphasis on biotic interdependence and its endemic relationship with the abiotic environment forces research to focus on permanent interactions in the biosphere.

Consistent with this paradigm, ecological ethics shares this epistemic approach, since it must break with the "Cartesian" linear sequentiality, which tended to impose an objectual, instrumental, and strategic rationality on relations with the natural environment. On the contrary, the rationality appropriate to ecological ethics is based on the integral valorization of the systemic relations between human beings and their natural environment. This shift can be perceived in the work of the American conservationist Aldo Leopold's "A Land Ethic" (2005) originally published as a chapter of *A Sand County Almanac* in 1949. This is the first work to explicitly use the category of ecological ethics. Its intention is to prescribe a new formulation of the Kantian categorical imperative to include the non-human members of the biotic community, understood by Leopold as "the Earth." The imperative he proposes is formulated thus: "A thing (or decision) is good when it tends to preserve the integrity, stability, and beauty of the biotic community. It is bad when it tends to the contrary" ([34]: 7).

The specificity of this approach lies in the fact that it proposes as a criterion of responsibility the protection of collective interests that go beyond the satisfaction of the exclusive needs of the human species. This extension of the categorical imperative to an interpretative framework that involves the environmental dimension leads us to consider as a determining factor the valorization of "ecological balances." This aspect presupposes that biological systems are in a dynamic equilibrium. Thus, a parametric change in one environmental variable tends to be compensated homeostatically, or by negative feedback, in another parameter.

This does not mean that these ecosystem balances are continuous and stable. On the contrary, biological systems manifest discontinuities, divergences, and tendencies to hysteresis, so that evolutionary history leads to situations different from the initial context. In this way, small changes in a biocenotic community can originate great divergences in the future based on non-linearity, in the feedback of variables, in the creation of cooperative effects, self-organizing phenomena, bifurcations, catastrophes and transitions to chaos ([51]: 152). This aspect, marked by uncertainty, radically challenges human responsibility, since it forces us to ethically face the systemic functional coercions of politics, law, and market economy ([6]: 133) that impact on the planet, without being able to have absolute certainties about the future effects of our present actions. Hence the relevance of Hans Jonas' prescription, regarding the need to adopt the principle of responsibility as a central imperative of all ecological ethics, which is synthesized as follows: "Act in such a way that the effects of your action are compatible with the permanence of an authentic human life on Earth" ([32]: 84–87).

3 Ecological Ethics in Pre-modern Societies

Although ecological ethics, explicitly marked with this name, only arises with the work of Leopold, a genealogical analysis can confirm that innumerable moral practices have been developed since time immemorial, which have rooted and defined the relations between human societies and the ecological communities into which they are introduced. It is a dimension inherent to human life and survival itself.

The concrete form of these moral practices lies at the level of social regulations, rooted in a particular *modus vivendi*, through traditional and intergenerational ties. Anthropology shows us that the "rules of use" of archaic societies, oriented to the regulation of human conduct, always considered ecological variables, given their predominantly adaptive functionality.

These traditional norms, often transmitted orally, accompanied by strong mechanisms of moral sanction, with high degrees of autonomy with respect to positive legislation, consolidated different models of "moral economy of the multitude" ([62]: 20–21). It was about "embedded" norms in social relations determined by an extra-economic rationality, different from the instrumental rationality ([43]: 117). In that context, the "world of life" was regulated by means of a sapiential ethical framework, of religious symbolic character.

Following Richard B. Braithwaite, we can understand religion as a heuristic concept that describes a moral system, based on stories that are added for illustrative purposes to encourage right action. But it is not necessary for subjects to *literally* believe the truth of those narratives for the stories to operate factually as guides to action; it is enough to "hold them to be true" [44]. In this way, the various religious traditions have built their moral systems aimed at ecologically adapting human communities to the environment, restricting and regulating their most diverse dimensions, including their productive, extractive, food, hygienic or medicinal activities, as well as their mobility and displacements, mercantile flows, care work, and reproductive tasks.

Examples of the ecological functionalities of religion can be reviewed in the works *Priests, Warriors, and Cattle: A Study in the Ecology of Religions* [35], and *An Ecological Interpretation of Religion* [38]. Also key works are *Theory of Cultural Change: The Methodology of Multilinear Evolution* [56], *Evolution and Ecology. Essay on Social Transformation* [57], and the voice *Ecology* in *Encyclopedia of Religion* [30]. Other relevant authors are Gottlieb [25], Hervieu-Léger [29], Reynolds and Tanner [47], Jones and Reynolds [33], and Burkert [9]. However, it is worth dwelling especially on *The Good Book of Human Nature* [63], which argues that forms of religious sociability developed to make sense of human transitions, in particular from an egalitarian hunter-gatherer ecological habitat of their own to a new habitat of agricultural societies. For van Schaik and Michel, religion emerges as an evolutionary strategy to cope with the unprecedented levels of epidemic disease, violence, inequality, and injustice that these human communities faced when they abandoned their nomadic-pastoralist way of life during the Neolithic revolution. The evolution of religious norms, tempestuous and conflictual, would reflect the need for ecological readjustment of human communities.

In this sense, it is worth noting the observation of Reyes Mate, who affirms: "In living religions there has been a millennial reflection on the forgiveness of the unforgivable, on the meaning of the meaningless life, or on the saving memory of the failed" ([46]: 155). Also noteworthy is the phrase of ([31]: 92) warning that Religion "in the good sense" is "the inexhaustible longing, sustained against reality, that this changes, that the banishment ends and justice arrives." It is the endless search for a good life, achieved when individuals manage to be "good" and "virtuous" in relation to the demands inherent in the social practice they execute ([36]: 15 ff.).

The binding character of ecological ethics with a religious matrix is based on the intimate perception of the subjects, who are constrained by bonds of obligatory reciprocity, and by mobilizing beliefs at the most intimate level of conscience. This subjective dimension is condensed in a series of convictions:

> Had we been intelligently conceived, to be controlled and recognized by a God who actively punished and rewarded intentions and behaviors, which would have helped reduce the frequency and intensity of our ancestors' immoral stumblings, and would have been, no doubt, favored by natural selection. ([8]: 21)

These customary forms of regulation had a strongly coercive character, since the sacred permeated everything, in a context where the division between immanence and transcendence was neither thinkable nor possible. The entrenchment of these shared religious certainties made traditional societies generate forms of production based on "common customs" which, along with prescribing procedures, interpreted the meaning of human activities. This is what the reading of Hesiod's *The Works and the Days of Hesiod*, which dictated a strict work ethic appropriate to its given context, produced in its primitive audience:

> The gods, indeed, withheld from men the sustenance of life; for otherwise, for a single day you would work enough for the whole year, living without doing anything. At once you would hang up the handle of the plough... and stop the work of the oxen and the patient mules. But Zeus concealed this secret (the secret of abundance), irritated in his heart because the shrewd Prometheus had deceived him. ([49]: 133)

These ecological ethics can be characterized by a number of characteristic features. They were material or substantialist ethics, since they sought to provide moral content in the form of shared goods, ends, intentions or values. On the ecological level, this is evidenced in a valorization of nature as a sacral realm, protected by sacred powers. For this reason, ecosystems were placed outside the field of mercantile exchanges, since the relationship with the ecological community prioritized use value over exchange value, assuming that nature was above all a relational, symbolic, and spiritual realm. Additionally, they were teleological ethics, since they prioritized the conviction in an ideal of fullness of life, which was expressed as "the good" or "the felicitous," locally situated, over a criterion based on "the just," understood as a universally demandable minimum.

At the same time, they were decisionist ethics, since they established the moral act as the result of an ultimate personal decision that could not be argued. This is explained by the fact that their form of justification was carried out through the interpretation of a symbolic universe in which the literal and the figurative were blurred, since the ultimate meaning of the norms always remained in a certain opacity for the rational mode of understanding. For this reason they were naturalistic ethics; they understood moral acts as empirically contrastable, based on a naturalized social order. These ethics succeeded in rigorously regulating life, but they were not "normative" in the contemporary sense, since they did not seek to logically ground their prescriptions for action. Finally, they were not consequentialist, because they were not justified by the outcome of the moral act, but emphasize attachment to permanent ideals of human flourishing, so they can be analyzed as virtue ethics.

4 Ecological Ethics After the Great Transformation

The loss of the binding power of the old ecological ethics, implicit in the moral systems of the traditional societies, has happened progressively, accompanying the processes of capitalist modernization. These changes have propitiated a "disenchantment of the world" ([66]: 231) that originated a new moral stage in the western societies, marked by the "polytheism of values":

> …life, insofar as it rests in itself and is understood by itself, knows only that eternal struggle between gods… (Life knows only) the impossibility of unifying the different points of view that, in the last analysis, can be held about life and, consequently, the impossibility of resolving the struggle between them and the need to opt for one or the other. ([66]: 225)

This effect, whose palpable expression is the desacralization of the structures and institutions of traditional morality, can only be explained by a much more fundamental change, which originates in the complete international reorganization of economic and social institutions. In this way, a profound transvaluation of human life takes place. Karl Polanyi calls this process "The Great Transformation," whose final result leads to the creation of new "fictitious commodities" (land, labor, and money), which in their original constitution had no commercial purpose. The result was that the land and the labor that worked it were radically dissociated. These ecosystems were reduced to their strictly economic function, which was only one of the many vital functions that they exercised in traditional societies. Thus, the value criterion that emphasized the ecological stability was lost, operating a substantial separation between the human being and nature ([43]: 238). This radical dissociation occurs to the extent that "market societies" are constituted, where the individual interest prevails:

> The real criticism we can make of the market society is not that it is based on economics - in a sense, every society, no matter what, must be based on it - but that its economy is based on self-interest. Such an organization of economic life is completely unnatural, which must be understood in the strictly empirical and exceptional sense. ([43]: 330)

For Macpherson [37] this process is based on the consolidation of the "possessive individualism" understood as the rational foundation of a "possessive market society," in which individuals are conceived as owners not only of goods, but also of their person and their work, understood as an alienable property.

Tawney uses a similar expression: "acquisitive society," which allows him to describe the *ethos* of modernity:

> Such societies may be called acquisitive societies because their whole tendency, interest and concern is to foster the acquisition of wealth ... It directs men not towards the fulfilment of social obligations - which limits their energy because it defines the goal to which it should be directed - but towards the exercise of the right to look after their own interests ... it assures men that there are no other ends than their own, no other law than their desires, no other limit than that which their own think advisable. Thus it makes the individual the centre of his own universe and dissolves moral principles into a choice between convenient things. And it greatly simplifies the problem of social life in complex communities, for it frees them from the activity of distinguishing between different kinds of economic activity and different sources of wealth, between entrepreneurship and avarice, between energy and unscrupulous greed, between legitimate property and property that amounts to theft, between the just enjoyment of the goods of labor and the idle parasitism of birth or fortune. ([61]: 33–34)

This change, of dramatic consequences in the area of ecological responsibility, does not seem possible to be reversed without simultaneously s altering the functional subsystems of present-day human society. It is not possible to retrace the path of modernity without also taking back the conquests of autonomy and self-determination achieved since the seventeenth century. Adela Cortina argues this point:

> It is said that in the Ancient and Middle Ages there was a great identification between each individual and his community, because they understood that the good of their community coincided with their own good. In that case it did not seem very necessary to justify the obligation to obey laws that benefited the community and also each of its members. But in the modern world the suspicion is introduced that the interests of an individual may not coincide with those of his community and, in such a case, why obey the rules? Modern versions of individualism will be an attempt to answer this question, which remains open today, because the dream of the "communitarians" of forming communities around an idea of the common good has not been realized. In our societies, on the contrary, individual and group interests come into conflict, and it seems difficult to go beyond that. ([14]: 63–64)

The formulation of proposals of ecological ethics, in the context of irreversibly pluralistic societies, must face at least three unavoidable tasks, according to Cortina ([16]: 1). First, to contribute to the clarification of what should be understood by ethics, in general terms. Next, to propose a foundation that offers a reasoned account of its specific moral proposal, arguing its justification and scope. Finally, to look for concrete applications, with a view to influencing civic life.

These three basic criteria, essential for any public ethics, constitute a lens through which to evaluate the different approaches in current ecological ethics. Through them, some of the main paradigms will be analyzed in a descriptive but also critical way, highlighting the fundamental aporias that run through these debates.

The critique of the approaches that will be detailed in the following sections does not imply disqualifying their contributions, undoubtedly relevant to the objective they propose. It is a criticism that presupposes, with Walter Benjamin, that "it is obliged to know by submerging" ([7]: 246). For that, it tries to show its internal limits in order to widen its current borders, allowing the urgent and necessary deployment of some essential ecological ethics.

5 The Weak Anthropocentrism of Bryan G. Norton

A first proposal to consider is that of Norton [42], based on his distinction between strong and weak anthropocentrism. His analysis tries to give foundation to a pragmatic approach, oriented to solve environmental problems, looking for an ecological ethics that is compatible with the ethical systems of Anglo-Saxon matrix, marked by individualistic rationalities.

His initial distinction lies in separating between anthropocentric and non-anthropocentric ethics. Norton assumes that only the human condition has intrinsic value and that humans are the only organisms capable of conferring value. Furthermore, he thinks that there are arguments for questioning the ontological commitments that non-anthropocentric approaches make by attributing intrinsic value to nature. In his perspective, environmental ethics should not assume in principle the demand to adhere to the non-anthropocentric paradigm, and therefore differs from the planning already exposed by Aldo Leopold, who sought to justify the attribution of intrinsic value to non-human natural entities.

To argue this point, Norton considers that there can be two forms of anthropocentrism, a weak and a strong one. By strong anthropocentrism he understands that where "*felt preferences*" prevail, which respond to any desire or need of a human individual that can be satisfied, at least temporarily, by some specific experience. In contrast, weak anthropocentrism prioritizes "considered preferences," understood as needs expressed after deliberation, which can be compatible with the global interest and have a hypothetical character. These are human desires, but expressed after careful deliberation where the consistency of these desires or needs is judged against a rationally adopted worldview. At the same time, he distinguishes between "habitual decisions," which affect individual equity, and non-habitual decisions, which he calls "allocation decisions," which concern the use of resources over a prolonged period of time. The latter would be the specific decisions of environmental ethics.

For Norton, weak anthropocentrism provides a sufficient basis for discerning individual and consumption needs, since considered preferences would be able to provide a critical basis for ecosystem-damaging value systems, generating universalizable obligations without the need to place intrinsic value on non-human natural entities. Norton's proposal of weak anthropocentrism provides a basic framework of environmental responsibility, assuming that it is impossible to live permanently in the total consistency that would require proceeding solely on the basis of rationally considered preferences. Ultimately, the ideal public citizen for Norton would be one who is guided by such preferences as much as possible. His proposal should be understood as the application to the field of environmental management of moral criteria based on the individual consideration of human needs and desires by means of systems of moral ideals, considered as universalizable rational criteria.

From this we can conclude that the main limitation of this approach lies in the fact that it goes back to the usual problems of "methodological solipsism," based on the *Faktum der Vernunft*, founded on an individual "I think." The monological features in Norton's ethics, in spite of all the nuances he makes to anthropocentrism, tend to legitimize an individualism that does not justify another rationality different from the teleological strategic one. This situation, called the "paradox of late liberalism" leads to a division of labor between positivism in the public sphere and decisionist existentialism in the private sphere. Norton falls into this paradox by reserving institutional decisions to the premises of a value-neutral procedural rationality and the decisions of the intimate sphere to individual valuations, under the assumption that integrating both poles would be an ideological attempt to restrict the freedom of the individual. Hence, Tom Regan considers that the anthropocentric approach should be considered more as an ethics for the administration or management of the environment and natural resources ("management ethics") than as a true ecological ethics ([45]: 19–34).

6 The Anti-ethics of Deep Ecology

The notion of deep ecology was proposed by Arne Naess in *The Shallow and the Deep, Long-Range Ecology Movement* [40]. His distinction between shallow (or short-range ecology) and deep *or long-range) ecology lies in the consideration of human beings on a plane of biocentric equality with ecosystems, which does not place them above or outside the environment. His critique of shallow ecology is that it would not go beyond the frameworks of anthropocentrism. Under this logic, all natural entities have the right to exist, regardless of their degree of consciousness or self-determination.

Naess assumes the approaches of fundamental interdependence between all natural phenomena, starting from Haeckel and converging with the studies of organicist biology, Gestalt psychology, quantum physics, the theory of relativity, and the mathematics of complexity. From there it follows, as a founding fact, that at both individual and societal levels we are totally dependent on the cyclical processes of ecosystems. In the face of this, Naess's anthropological proposal is described as a process of ecosophical "self-realization" [41], where the maturation of individual self-consciousness leads to a progressive identification with all living beings, human and non-human. It is a project of overcoming the binary dualism subject/object, which ends up disappearing to the extent that the borders of the individual "I" give way, until experiencing the empathic unity with all living things, and the full awareness of their interrelationships.

With the individual self dissolved, an "Ecological I" takes its place, intimately connected to all living beings. In this way, the notion of moral duty disappears (insofar as self-determination coincides with the expression of a consciously shared destiny) in symbiotic unity with the ecosystem. To the extent that the self and the World coincide in a transpersonal process, the moral self becomes an illusion that vanishes before the full consciousness of the *continuum* that unifies the human being and the non-human entities with which we form "knots in the biospheric web."

Villarroel observes that this treatment of the "Moral I" does not consider that "from a scientifically ecological perspective the world is revealed as a differentiated and not homogeneous unit of the natural" ([64]:196). The biokenotic systems show an accumulation of "discontinuous functions" in which a determined variable suffers a jump or sudden change of value. Moreover, discontinuity appears as the specific property of the discrete states of matter, for which ecosystems operate under the principle of differentiation of their singular stable elements, provided with qualitatively different and determined structures. In this way, ecological experience manifests itself rather as a permanent dialectic of discontinuity and continuity, in the field of spatio-temporal properties and relations, and of movement and interconnection between biotic and abiotic entities.

This dialectical aspect is subsumed by Naess's ecosophical self-realization proposal which, by dissolving the moral self into a transpersonal "Ecological Self," prevents the recognition of the discontinuity existing between human beings, biotic entities and abiotic entities. The affirmation of continuity prevents assuming the need of ethical responsabilization of a subject, individual or collective, that can never dissolve its otherness in differentiated entitative realities. For that reason, Villarroel ([64]: 195) maintains that this aspect of deep ecology becomes a form of antiethics, to the extent in which the symbiosis I/world prevents the assumption of the differentiated responsibilities that concern human beings and their institutions in front of ecosystems.

7 Ecofeminist Ethics

To understand ecofeminism, we need to understand a set of approaches that have gravitated in social movements since the late 1970s. In part, these divergences in ecofeminism are related to the general debates in the feminist movement, which move between radical, liberal, and socialist positions, and the influences of the peace movement. A tension exists between a spiritual or culturalist ecofeminism, which tends to highlight the "natural" affinity of women with ecosystems, and a socialist ecofeminism that is oriented by a political approach. The latter highlights that the links between women and the environment are determined by a complex structure of domination that includes class, caste/race, and gender variables, and which organizes production, reproduction and income distribution according to a patriarchal interest.

The spiritual or culturalist current of ecofeminism tends to affirm essentialist positions, insofar as it prescribes a "feminine principle" ([52]: 26) that would serve as an ethical criterion of life and the basis for sustainable development. This principle presupposes that women have a role as "innate caretakers of the planet" as they are the direct victims of environmental degradation. The feminine principle would express an ethic of care, as women and nature function in a role of "letting grow and making grow" ([52]: 67). This principle highlights the existence of a differential nature between men and women, which would place the feminine gender as naturally closer to ecosystems and, therefore, endowed with greater proclivity and sensitivity to environmental dilemmas and conflicts. To this end, it highlights the historical, biological, and social connections between women and the natural environment, and assumes that the patriarchal order is the common cause of the exploitation and oppression of women and nature. The application of the feminine principle would contribute to a "survival perspective" that would restore a harmonious relationship between environment and human society.

The critique of spiritual or culturalist ecofeminism points to the direct relationship established on the basis of the feminine principle between women and nature, rooted in biological factors. In this way, this ecofeminism becomes essentialist in a form of naturalist ethics that identifies the phenomenal experience with what is morally appropriate by assuming that moral judgments have a merely descriptive character, which has been called "naturalist fallacy." This occurs when a descriptive statement of an empirical fact (for example, a phenomenological observation that attributes a greater willingness of women to be concerned about environmental problems) becomes a value judgment. For socialist ecofeminists, it is necessary to notice the differences in women's life conditions, which never become a homogeneous group:

> 1) it postulates women as a unitary category and does not differentiate women according to their class, race, ethnicity, among other factors; 2) it locates the domination of women and nature almost exclusively in the field of ideology, ignoring the material sources of this domination; 3) even in the field of ideological creations, it says little about the social, economic and political structures within which these creations are produced and transformed; 4) feminist reasoning does not consider the relationship that women experience with nature

> as opposed to the relationship that they or others may conceive; (5) currents of ecofeminism that attribute the connection between women and nature to the biological can be seen as adhering to a form of essentialism. This formulation disappears in the face of the ample evidence that the concepts of nature, culture, gender, etc., have been historically and socially constructed and vary from one culture to another, within the same culture and from one epoch to another. ([2]: 244)

This criticism is also shared by ecofeminist Warren ([65]: 24), since the affirmation that certain ethical preconditions are given by nature entails a link to naturalistic fallacies, such as those that affirm the "natural" generosity of women, which hides forms of factual oppression linked to the malleability of their condition of subordination.

8 Utilitarian Extensionism

A relevant approach within current ecological ethics are the "extensionist" proposals, which have in common the search for the recognition of an intrinsic value in non-human living beings. These elaborations began in the 70 s, when a debate began in moral philosophy regarding the moral status of "non-human animals" that, like our species, possess *sentience* (in the utilitarian perspective) and, therefore, should be given necessary consideration.

A key role in this discussion has been played by the social mobilization that in recent years has questioned the treatment that humans have given to other species, especially through the industrialization and technification of meat production. But the academic framework of the problem took shape with three innovative works: *Animals, Humans and Morality* [24], *Animal Liberation* [53] and *The Moral Status of Animals* [10]. The premises put forward in these texts demanded a radical change in the human attitude towards nonhuman animals, as the authors demanded an end to the normalization of the exploitation of other species as an inevitable and morally neutral reality. On the contrary, it claimed this consideration a "continuous outrage" ([54]: 7).

The publication of these books complemented the development of an incipient applied ethics, aimed at concretizing the debates that moral philosophy had tended to direct to spheres of great abstraction. Since the publication of these works, a growing number of philosophers have spoken out on the problem of the moral status of nonhuman animals, including Paola Cavalieri, Jacques Derrida, Raymond Corbey, David Livingstone Smith, Adela Cortina, Matthew Callarco, Cary Wolfe, Will Kymlicka, and Kari Weil. In them there is a common pattern, which points to the assumption of demands for the recognition of a specific intrinsic value or inherent dignity in living, non-human beings. However, they differ in other aspects, such as the recognition of "rights" of some species, and the hypothesis of morally homologating human and non-human beings on the level of their intrinsic value.

The argument of extensionism lies in identifying a contradiction in the Western moral tradition that has assumed the "principle of reason" as the basis for the attribution of moral competence. For example, from an anthropocentric deontological framework, animals have only a relative value, as means, while human beings are to be considered ends in themselves. Thus, the human species should be considered as different in rank and dignity from irrational animals, which could therefore be treated with discretion or, at least, as objects of care and compassion, mainly because of their attachment to human beings. For the contractualist view, the exclusion of other species is based on the fact that only beings with the capacity to contract can be subjects of law and be part of the moral community. Human beings can contractually define certain obligations with respect to ecological systems and other living beings, but they do not act as signatories of such contracts. Hence, there would be no direct moral duties in relation to animals, because their value would be only instrumental and because they do not have the capacity to be part of communities based on norms constructed by themselves.

Extensionism criticizes these assumptions, arguing that under them some human beings, such as young children, the elderly in profound senility, the temporarily unconscious, or the mentally handicapped, could not be considered morally. On the other hand, some non-human species could be incorporated in this field, since science has shown that they possess highly developed communication capacities such that the standards of morality should be extended to assume their specificity. To resolve this contradiction, Singer appeals to the criterion proposed by Jeremy Bentham that incorporates *sensitivity*—understood as the capacity to feel pleasure or pain—in the criterion of moral inclusion, since this homologates human beings independently of their capacity for reason and communication, and assumes the consideration of non-human animal beings that possess this competence.

The moral task for utilitarian extensionism is to maximize, as far as possible, the interests of all sentient beings. This must include the interests of animals in the scope of moral consideration, assuming that their natural interest is to avoid pain and suffering. Hare ([28]: 219) formulates this principle as an extension of the golden rule, which should be applied to all sentient beings in the measure of their own nature. For that, a utilitarian calculation would be necessary, where the moral judgement is based on weighing the damage caused:

> If there is no harm, there is no evil. Moreover, it has to be globally harmful. If a course of action involves some harm, but greater benefits, and there is no alternative with a greater balance of good over harm, it will not be evil. We have to ask, therefore, whether the whole process of raising animals and then killing them for food causes them more overall harm than good. My answer is that, assuming, as we must assume if we want to keep the "killing" argument distinct from the "suffering" argument, that if they are happy while they live, it does not. For it is better for an animal to have a happy life, however short, than no life at all. ([28]: 226)

Faced with the arguments of extensionism, it is worth stopping at some objections that lie in the crucial distinction between the necessary attribution of value to animals and the recognition of rights to non-human living beings. In this regard,

Cortina [18, 19] proposes some distinctions that start by considering that the process of establishing moral norms is essentially intersubjective. People possess the capacities of communicative competence necessary to deliberate and dialogue on social moral norms through the use of reason, self-awareness, dignity and reciprocal recognition, which constitute preconditions of the communicative competence of moral discourse.

This capacity is a presupposition prior to social dialogue; it is an inalienable and non-negotiable condition from which human rights are founded as absolute prerequisites for moral deliberation. Non-human living beings, not possessing these capacities, do not form part of the ethical discursive community, and therefore do not have rights in the strict sense, prior to any subsequent moral consideration. Hence every right is, in the end, a human right. However, non-human living beings have by nature an inherent or internal value that is necessary to take into account in the moral dialogue of society, which must enact moral obligations of care and responsibility to protect that value ([18]: 62). From there arises the distinction between granted rights and recognized rights:

> Do animals have rights? The answer could not be more obvious today: yes, of course, they have the rights granted to them by the laws of a good number of countries, which increasingly specify the treatment that should be given to animals; a treatment that, at the very least, requires not causing useless suffering … However, the question "do animals have rights" usually refers to a more complicated question: whether they have a type of rights similar to human rights, which are not granted, but must be recognized. Human rights are prior to the will of legislators and oblige them to recognize and embody them in concrete legislation. It is not the same thing to grant a right, which may or may not be done, as to have to recognize it. We have a lot at stake in this difference. [20]

This point of view realizes that a casuistic and material approach, such as the recognition of inherent value that accounts for a dignity, is sufficient to relativize human needs and re-evaluate them in relation to the capabilities of non-human animals, and thus safeguard the notion of rights for moral subjects capable of rationally recognizing the transcendental sense of the moral imperative ([18]:172).

9 Biocentric Ethics

While Peter Singer, Stephen R.L. Clark, and S. Goldovitch inscribe themselves in an extensionist approach, which is defined as "an extended ethics that draws the line of the division of moral consideration around all sentient creatures, but leaves other beings outside the line" ([54]: 345), other authors, such as Tom [45] and Bernard [48] propose a "biocentric" ethics that does not admit this distinction. Unlike the extensionist utilitarianism, biocentrism does not base the attribution of moral rights to non-human species on the basis of their capacity to feel pain and pleasure, and, therefore, it does not make a calculation of utility. For biocentrism, the mere attribution of life to subjects would be enough to assign basic moral rights and an inherent value, as patients and moral agents, to human and non-human beings alike on an equal footing. Taylor ([60]: 172) considers that living organisms not only possess individual bio-physiological identities, but also individual personalities, for which they must be recognized as centers of teleological action. This purpose could be synthesized in the permanent tendency of species to maintain themselves in existence, as well as to deploy the biological operations that allow them to reproduce and adapt to the different events and environmental contexts. The right of all life is to be able to deploy its capacities and conditions of flourishing within what each species defines as a good life. In this way, the biocentric imperative would translate into not using other beings, under any case, circumstance, or superimposed valuation, by virtue of their teleological realization and specific flourishing.

However, biocentric ethics poses serious difficulties at the moment of defining methods that allow one to solve the daily moral dilemmas linked to the valuation of non-human living beings, since the substantivization of the *telos* of each species is formulated from a place totally independent of human valuations. This deficit is produced in the material and concrete moment of ethics, which consists in giving content to the formal structure of a given ethical proposal.

This approach begins by recognizing that all ethical experience starts from an inescapable "epistemic anthropocentrism" as a source and assumption of all moral deliberation. Being what we are, as a species we perceive and conceive the world, sensorially and neuronally, from an anthropocentric perspective from which we cannot escape. Based on this recognition, human complexity can establish a priority before a less complex organism. In that way, it would not be unjust to use other non-human animals as a means to our ends, if by doing so we do not do violence to their specific *telos*, understood as the flourishing of their essential capacities. On the contrary, what would be unjust would lie in contradicting the animal's *telos* and frustrating its possibilities of leading the good life characteristic of its species.

10 Latin American Social Ecology

In Latin America, the contribution of social ecology has acquired social and political importance, especially through the work of authors such Gudynas and Evia [26], Acosta [1], Seoane [50], and Svampa [58, 59]. This approach assumes a paradigm inherited from "geonomism," which at the beginning of the twentieth century had already considered that the problems of economy, climate, and environment are linked; therefore, the solutions to these problems must also be linked. In the social ecology of Gudynas and Evia, it is recognized that the environment is a heterogeneous system, subject to a permanent hermeneutic of the actors that interact through very diverse symbolic frameworks. Human-environmental relations reflect this field of interpretations. Social ecology must study these processes, from an ethically committed praxis, which the authors describe as "an explicit ethical posture in favor of all forms of life" ([26]: 32). In this way, Gudynas and Evia formulate the following methodological axioms on which they base their work:

> a) Environmental systems cannot be studied in isolation from human systems, from which there should not be a social science different from an ecological science; b) Consequently, there is no social practice that is different from an environmental practice. Environmental work is, at the same time, social work; c) The dynamics of processes must be recovered, which demands emphasis on the histories of both human beings and environments; d) As any person or group reaches particular interactions with its environment, the study of those relations must necessarily be done with the participation of the people involved. ([26], 31)

This analytical and procedural methodology is based on a series of ethical axioms coherent with the epistemological approach already exposed:

> a) Intrinsic values are recognized in other living and non-living things; (b) The ways in which any person knows and feels the environment with which he or she interacts must be respected; (c) The results of research are ethically neutral, but not in their starting points, the process of research, or the use of those results. ([26], 31)

Finally, from these axioms follow principles about the application of knowledge:

> a) Technology must be on a human scale, promoting human development, and in accordance with the dimension and objective of the problem being faced; b) It must also be socially and ecologically sustainable and equitable, recognizing values in the environment. ([26], 31)

Based on these axioms, Latin American social ecology has provided a systematic critique of economic models based on "extractivism," a category it defines as "a particular type of extraction of natural resources, in large volume or high intensity, of which 50% or more is destined for export, as raw materials without processing or with minimal processing. It includes both the phases of exploitation and the previous phases of exploration, discovery, etc., and also the subsequent phases, such as closure and abandonment of the appropriation sites" ([27]: 15). He distinguishes between a "classic extractivism" implemented by transnational companies while the state assumes a secondary role, providing protection for investments and exports with low levels of taxation, and a "neo-extractivism" where there is greater state presence through state or mixed companies, with a higher level of royalties and even greater controls over production. Recognizing that this neo-extractivism allows States to collect more taxes, which helps to implement new social programs, social ecology emphasizes that the financing of all these programs continues to be based on a conventional model, under a pattern subordinated to the export of natural resources, especially minerals, hydrocarbons, and monocultures.

This situation makes states dependent on these processes, caught between popular demands to maintain and increase funding for social programs and, on the other hand, pressured by protests linked to the advance of mining and oil companies in communities that are affected by serious pollution and forced displacement, even to the point of assassinations of local leaders, especially in indigenous communities and tropical zones. To describe this specific context, Gudynas has proposed the concept of "extrahección," which he constructs from the Latin *extrahere*, "where 'ex' means outside, and 'trahere' alludes to removing and dragging towards oneself" ([27]: 11). Therefore, he defines *extraher* as the most acute case of appropriation of natural resources, where these are extracted through violence, and human and Nature's rights are violated. "It is not a consequence of a type of extraction, but is a necessary condition to be able to carry out the appropriation of natural resources" ([27]: 15). As an alternative, social ecology focuses on the promotion of the paradigm of *Good Living*, which in cases including Bolivia and Ecuador, has constitutional status. This proposal involves implementing a post-extractivist economy aimed at overcoming the contradictions of the current model.

The criticisms of social ecology lie at the pragmatic political level, since a transition to a post-extractivist economy could only be generated through a gradual program of productive transitions on small and medium scales, which through the visibility of concrete and viable cases, would generate a base of social and political support that would promote changes in the macro scale of production. For this reason, neo-extractivism argues that the resources generated by the current system can leverage this transition, guaranteeing employment, and social and environmental policies that today can be implemented through a State that has more funds for these objectives.

From this perspective, Latin American social ecology would show a deficit at the level of its pragmatic mediations, since between the denunciation of neo-extractivism and the possibilities of implementing the paradigm of "Good Living" there would be a problem of political and temporal scale. This discussion has been reflected in the so-called "Gudynas-Harvey debate," which has highlighted that the critique of neo-extractivism has a problem in the analysis of the global dimensions of capitalism, especially international power relations. Therefore, the processes of extrahección, being absolutely rejectable, presuppose a complex interpretative framework of political relations that exceed the local and regional, and require a less descriptive, abstract, and unidimensional approach [39]. As Apel warns, "the morally restrictive functional coercions (of the right and of the market economy) are not abolished with an ideological critique and are not reduced to conflicts of norms of interpersonal action" ([6]: 146).

Another way of approaching the critique of Latin American social ecology is formulated by García Linera [22], who leads the debate to a re-edition of the Weberian dichotomy between the "ethics of conviction" and the "ethics of responsibility" in politics. It should be remembered that Weber addressed this problem in his lecture *Politik als Beruf* (2003), which placed two variables in tension: the rationalization of society requires specialized professionals in power and government, but at the same time politics requires commitments that go beyond what is strictly stipulated in a legal or technical norm. That is why Weber uses the German word *Beruf*, which can be translated into English as both *profession* and *vocation*. The ambiguity of the German term comes to suggest that politics is always a "specialized work," as well a "vocational" activity, understood as a response to a "call" that involves something more than a paid remuneration. This double aspect stems from the social purposes of politics, which give it its meaning and legitimacy.

The need for this vocational dimension in politics is unique. It is related to the contact and use of power—a context that demands discernment at every moment about possible pacts with necessary but corrupting "diabolical powers." In this sense, politics, as a specialized professional activity, would not be suitable for everyone. The Mephistophelian dimension of politics explains that technical specialization is not enough for its exercise. The political practice demands a capacity to search and identify the internal ends and goods of this activity, assuming that the exercise of power implies an insoluble contradiction in the plane of its pragmatics. Under these assumptions, García Linera and other political leaders attribute to the Latin American social ecology an incapacity to process its own political tensions, maintaining itself as an abstract technical analysis, which they qualify as "political faith decalogues," and which do not assume the complexity of compatibilizing conviction and responsibility at the moment of facing the systemic functional coercions of the market economy. The principle of political responsibility obliges ecological ethics to discern a complex dilemma of inescapable character: neo-extractivism, as a mechanism for satisfying needs, is demanded by those who need it for their functional subsistence, despite the fact that at the same time these same subjects are direct or indirect victims of the violent and destructive dimensions of this productive model.

11 Conclusion and Proposed Rationale

The brief review of these different models and ecological ethical proposals has shown the specificities, contributions, and limits of the analyzed approaches. It is not possible for us to fully explore the study of each field of authors, so we have chosen to stop at their critical cores, which determine their conditions of applicability.

The ecological ethics of substantialist and teleological matrix, linked to religious traditions, to the adaptive frameworks of traditional societies, or those proposed by contemporary authors such as Arne Naess, Vandana Shiva or Tom Regan, have an enormous binding capacity for those who adhere to them. Particularly relevant in these approaches is their profound sense of the sacredness of nature as *res extra commercium,* impossible to commodify and reduce to its utilitarian dimension.

However, the limitations of these approximations lie in the impossibility of overcoming the frameworks of "agathological maximalism" ([15]: 32), a conception of the good founded on convictions rooted in the intimate sphere, on the basis of pre-rational valuations. If ecological ethics is reduced to this dimension, the logical consequence is the imposition in the political ethical plane of the paradox of late liberalism, which leads to a division of work between positivism in the public sphere and decisionist existentialism in private life. Objectivity is assumed as an exclusive predicate of techno-science, while the normative prescriptions of ethics are classified in the field of "subjectivist-irrationalist existentialism." Thus, the field of institutional economic decisions should operate exclusively under the premises of a value-neutral procedural rationality, and individual life under the primacy of subjective and inalienable convictions. Therefore, the foundation of an ecological ethics in a post-metaphysical context, after the "Great Transformation" described by Polanyi, should aspire to conditions of legitimacy in the face of a complex and plural society. The challenge is to lay the foundations for an ecological ethics adequate to valuational pluralism, which allows the integration of persons, peoples, and cultures with different interests and moral traditions, coming from different vital worlds ([4]: 93).

Another aspect to take into account lies in the applicability of the ethical proposals, articulating the necessary ethical–political tension between conviction and responsibility. A first criterion to resolve this point lies in prioritizing the "principle of responsibility" formulated by Jonas ([32], 84–87), which mandates the criterion of "generational interdependence." In this way, ethical responsibility presupposes some duties of the present generation towards future generations, but also requires attending to the needs of the present generation, especially of those who live in conditions of poverty that put their dignity and subsistence possibilities at risk. Subject to this criterion, neither the utilitarian proposal of Singer nor the ethical-ideological critique of Gudynas and Acosta are sufficient.

Overcoming these aporias requires a framework of meta-ethical foundations that allows the conditions of validation of ecological ethics, as applied ethics, to be broadened. Taking into account the problems identified, any ecological ethics proposed should have the following criteria:

A. It should be a normative ethic, since it should be able to inform practical orientations for action at the intersubjective and institutional level, not just a teleologically inspired framework for individuals.
B. But this normative foundation cannot be naturalistic, since it should understand that its moral predicates are not reducible to empirical facts. On the contrary, it must be a cognitivist ethics, since it should consider that its moral judgments can be argued by means of rationally grounded intersubjective agreements.
C. In this way it would be a formal ethics, based on procedural criteria, which have as their horizon the possibility of universalization.
D. In this context, an ethics applied to the ecological field should be an ethics of responsibility of a deontological nature, since it should prioritize duty before "what is just," understood as what is universally demandable, which is constituted as that to which we are obliged before others.

In coherence with this analysis, it is valuable to review the contribution of Karl-Otto Apel, especially his work related to the application of discourse ethics, based on the universal pragmatics of language. This philosopher distinguishes two lines in the development of his discourse ethics ([5]: 160): a part A, abstract-ideal, which includes the ultimate pragmatic-transcendental foundation of general and situational norms, and a part B, that looks for the conditions of application of the procedural fundamental norms of part A by means of an "ethics of the solidary responsibility referred to history" ([6]: 310). The interaction between parts A and part B allows a mediation between consensual rational action and strategic rational action ([5]: 175). This mediation operates by means of the regulative criterion of the "ideal community of argumentation" ([5]: 21).

Apel proposes an "ethical-discursive reserve" that allows a framework of responsibility before "the autopoiesis of institutions, of objective coercions and of functional systems" ([5]: 149). Faced with the self-reference of politics and economy, the pretensions of ethical-discursive validity must assume a universalist point of view, but different from that which emerges from the neutrality regarding the values of scientism. This universalization of the criterion goes through assuming a pragmatic-transcendental principle oriented to the adoption of norms, and based on criteria of justice, which Apel formulates in the following way:

> All beings capable of linguistic communication must be recognized as persons, since in all their actions and expressions they are virtual interlocutors, and the unlimited justification of thought cannot renounce any interlocutor and any of his virtual contributions to the discussion. ([3]: 380–381)

This universalization principle is complemented by Apel with a principle of "the action" or of "complementation" that allows to overcome the existing separation between the discursive ethical rationality and the strategic rationality, and also the separation between the deontological ethics and the teleological ethics ([11]: 260). The principle of complementation relates to the formal maxim of collaborating in the realization of the conditions of application of the principle of universalization in specific contextual situations. It is a mechanism proper to the self-reflexivity of discourse, therefore, it is the teleological complement to the deontological principle. These principles start from the recognition of all participants as equally free and rational persons, which founds a "human responsibility for the historical creation and the necessary changes of social institutions" ([6]: 195). It allows to establish a "solidary responsibility for the consequences and subconsequences of human activities" ([5]: 148). It deals with the possibility of assuming the secondary effects of collective actions and activities, incorporating the co-participants who cannot intervene directly or cannot follow the procedural rules in the communicative interaction.

This framework allows for the foundation of a universalist ecological ethic that assumes the inevitable confrontation between strategic rationality (proper to economic and political institutions) and the communicative consensual rationality that emerges from the world of life based on the demands of respect for the dignity of ecosystems and biocenotic communities. The process should not lead to an overlapping consensus, dependent on the factual consensus among those affected who today can intervene in the deliberation. What Apel proposes, by means of a regulative principle of universalization, does not only concern those who have been formally present in the formation of the factual consensus existing today. What he proposes, as a pragmatic counterfactual criterion, is an "unrestricted consensus" that makes it possible "to initiate and regulate a change of the political conditions of the formation of consensus in order to take into account the interests of all the affected parties" ([6]: 185).

From the previous foundation we can deduce the Apelian distinction between the "real community" and the "ideal community of communication." The real community lies in the facticity of the sociocultural forms of life ([5]: 164). In it, language constitutes a network of reciprocal recognition that provides the basis of contingent identity of people and institutions. In turn, the ideal community is a regulative idea and, therefore, a "counterfactually anticipated community" ([5]: 157) in which "the opportunities to choose and carry out speech acts are symmetrically distributed, and it is guaranteed that the roles of dialogue are interchangeable" ([12]: 130). It can be considered a teleological principle, since it proposes that the validity of norms is not decided by the peculiar intuition of each individual, but that personal decisions on normative validity are contrasted, by means of procedural rules, with their possible acceptance in an ideal community of argumentation ([13]: 78). In this way, it operates as a regulative principle that consecrates equal rights and equal co-responsibilities.

The real community has preeminence, since the ethics of discourse takes into consideration the hermeneutics of the facticity and historicity of being-in-the-world, and the necessary belonging to a socioculturally determined "form of life" ([5]: 166). In this sense, the ethics of discourse responsibly assumes the contextual consequences of its application, which cannot be made to depend exclusively on an unconditionally valid moral principle. Therefore, the "part B" comes to contribute a strategic mediation that is focused on two principles:

(1) the preservation of the real community, understood as the subject of discourse and all those who depend on it;
(2) to lay the material and cultural foundations of the ideal community of communication, without endangering its own and others' preservation ([21]: 157).

Principle 1 constitutes the major premise necessary for the achievement of principle 2, but at the same time principle 2 confers meaning to the principle of conservation affirmed in 1 ([11]: 262). Thus, premise [2] can operate as that "restrictive condition" that prevents putting at risk the previous conquests of humanity (the democratic state of law and constitution) in function of a determined future. This distinction introduces a meta-criterion that allows the interweaving of reality and ideality: the survival of the real community without this limiting the attempt to prefigure the ideal community in the real community ([3]: 409). Not only is the foundation by consensus possible, but also:

> One must also take into account the fundamental rule of history-related responsibility - that of the cure (*Sorge*) - for the preservation of the natural conditions of life and of the historical-cultural achievements of the actual community of communication existing, in fact, now. ([5]: 165)

Responsibility and solidarity are integrated in this way insofar as "solidarity responsibility" is recognized as a transcendental condition in intersubjective communication. This interweaving accounts for the application of ethics that depends factually on the direct cooperation of the persons involved or their representatives. It is possible that those involved refuse to cooperate, opting for overt or covert conflict. Faced with this possibility, Apel warns that it is possible to apply the institutionalized system in the current legality, insofar as positive law can act as a "complement to morality," especially in those cases in which a lasting and binding resolution of conflicts is peremptory.

This consideration of strategic rationality in part B, manifested in the recourse to the rule of law, presupposes that the conditions of application of the ethics of discourse are and will be imperfect. In this way, the preservation of the real community acts as a principle of responsibility, since there is no certainty of the good will of the other participants in the deliberative process. This is not meant to confuse the criteria of validity that originate from discursive ethics and juridical duties in law. While the moral norm is unconditional and governs independently of its legal form, the legal norm has mechanisms of sanction and political institutionalization. Similarly, democracy, as a guarantee of equality in political participation, leads to the concretion of principle D, as made explicit by Apel, quoted by Adela Cortina:

> Democracy as an approximation to this ideal requirement is, therefore, something more than a set of neutral procedures that we decide to follow for pragmatic reasons; it has its ethical-normative foundation in the ethics of the ideal community of communication that we have always accepted when arguing. ([11]: 181)

This explains that part B of Apelian discourse ethics can provide the foundational criteria for an ecological ethics that overcomes a series of current aporias. This does not obviate the need to complement this contribution with specific mediations in the field of social sciences, law, politics, or economics, which allow its application at the micro, meso or macro institutional level.

A specific aspect that a dialogically grounded ecological ethics must consider is the dimension that escapes mere rational intellection. In this field there is room for the "ungraspable silence of nature." That silence can become "common understanding," if it is possible to find a "common feeling" from a communication that integrates the vital experiences by means of a "Principle of cordial recognition" ([19]: 205–216). This principle comes to extend responsibility "up to where it reaches the limit of intersubjectivity," beyond the borders of humanity, until incorporating the recognition of all "beings worthy of respect and care." From there, Cortina understands that recognition can be operationalized from the following civic principles:

> 1. Principle of non-instrumentalization of people; 2. Principle of capabilities; 3. Principle of distributive justice; 4. Principle of dialogic principle; 5. Principle of responsibility for defenseless non-human beings. ([19]: 222 ss.)

These principles allow us to understand that the capacity to communicate goes beyond the argumentative dimension. For this reason, the "silence of nature" is also a phrase that interpellates when asking for an account of non-compliances, confronting those who have evaded a responsibility or a contracted duty, or recriminating for an act of injustice. Intersubjective interaction goes far beyond logical argumentation, since it unfolds with the very existence of the interlocutor. In this way, the ethics of discourse is not reduced to its purely procedural or instrumental dimension. The criteria of normative validity oblige the interlocutors to commit themselves to "work actively to raise the material and cultural level of those affected, so that they themselves can defend their interests" ([17]: 120). Relationships of co-responsibility and commitment are required, which must translate into the empowerment of the affected peoples', without which any dialogue is impossible ([17]: 20). These principles, Cortina maintains, allow a more complete understanding of the communicative link, not only as argumentation, but as an estimative, interpretative, and appreciative capacity of recognition and sense of justice. Adela Cortina shows that the communicative bond presupposes the capacity to achieve "cordial recognition" and "compassionate recognition." These experiential aspects are those that allow "the unfolding of the potentialities of the communicative bond" ([17]: 124).

References

1. Acosta A (2011) Extractivism and neo-extractivism: two faces of the same curse. In: Grupo permanente de trabajo sobre alternativas al desarrollo (eds) *Más allá del desarrollo*. Abya Yala/ Fundación Rosa Luxemburg, Quito, pp 83–118
2. Agarwal B (1998) El debate sobre género y medio ambiente: lecciones de la India. In: Vázquez García V, Velázquez Gutiérrez M (coords.) Miradas al futuro. PUEG, CRIM, CP, pp 239–285, Mexico
3. Apel K-O (1985) The transformation of philosophy (vol II, The apriori of the community of communication). Taurus, Madrid
4. Apel K-O (1986) Estudios éticos. Alfa, Barcelona
5. Apel K-O (1991) Teoría de la verdad y ética del discurso (Introduction by A. Cortina). Paidós, ICE-UAB, Barcelona
6. Apel K-O (2007) Globalization and an ethics of responsibility. Prometheus, Buenos Aires
7. Benjamin W (2007) Presentation of the journal 'Angelus Novus'. In: Obras (Libro II/vol 1). Abada, Madrid
8. Bering J (2011) The instinct to believe. Paidós, Barcelona
9. Burkert W (1996) Creation of the sacred. Tracks of biology in early religions. Harvard University Press, Harvard
10. Clark SRL (1977) The moral status of animals. Clarendon Press, Oxford
11. Cortina A (1985) Razón comunicativa y responsabilidad solidaria: ética y política en K.O. Apel. Sígueme, Salamanca
12. Cortina A (1986) Ética mínima. Tecnos, Madrid

13. Cortina A (1993) Ética aplicada y democracia radical. Tecnos, Madrid
14. Cortina A (1998) Hasta un pueblo de demonios. Ética pública y sociedad. Taurus, Madrid
15. Cortina A (1999) Citizens of the world. towards a theory of citizenship. Alianza, Madrid
16. Cortina A (2003) Lecture: moral pluralism, ethics of minimums and ethics of maximums. In: Cyber humanitatis, [S.l.], n. 27. http://www.cyberhumanitatis.uchile.cl/index.php/RCH/article/view/5706/5574. Accessed 30 May 2017
17. Cortina A (2007) Ética cordis. Isegoría 37:113–126
18. Cortina A (2009) The frontiers of the person. The value of animals, the dignity of humans. Taurus, Madrid
19. Cortina A (2009) Ética de la razón cordial. Educating citizenship in the 21st century. Nobel, Oviedo
20. Cortina A (2010) Do animals have rights? In: El País. http://elpais.com/diario/2010/07/29/opinion/1280354405_850215.html Accessed 15 Jun. 2017
21. Cortina A, Martínez Navarro E (2001) Ética. Akal, Madrid
22. García Linera A (2011) The "Oenegism." Childhood disease of right-wingism. Vicepresidencia del Estado, La Paz
23. Garrido Peña F (2007) On ecological epistemology. In: Garrido Peña F et al. (eds) El paradigma ecológico en las ciencias sociales. Icaria, Barcelona, pp. 31–54
24. Goldovitch R, Harris J (1972) Animals, men and morals. Taplinger Publishing Company, New York
25. Gottlieb RS (ed) (2006) The Oxford handbook of religion and ecology. Oxford University Press, Oxford
26. Gudynas E, Evia G (1993) Ecología social, Manual de Metodologías para Educadores Populares. Editorial Popular, Madrid
27. Gudynas E (2013) Extractions, extractivisms and extrahecciones. A conceptual framework on the appropriation of natural resources. Observatorio del desarrollo, CLAES, Montevideo
28. Hare RM (1993) Why I am only a demi-vegetarian. In: Essays on bioethics (chap 15). Clarendon Press, Oxford
29. Hervieu-Léger D (1993) Religion et écologie. Le Cerf, Paris
30. Hultkranz A (1966) Ecology. In: Eliade M (ed) Encyclopedia of religion. University of Washington Press, Washington, pp 581–585
31. Horkheimer M (1974) Notizen 1950 bis 1969. S. Fisher Verlag, Frankfurt
32. Jonas H (1995) El principio de responsabilidad: Ensayo de una ética para la civilización tecnológica. Herder, Barcelona
33. Jones E, Reynolds V (eds) (1995) Survival and religion: biological evolution and culture change. Wiley, New York
34. Leopold A (2005) Una ética de la tierra, 2nd edn. Catarata, Madrid
35. Lincoln B (1991) Sacerdotes, guerreros y ganado: un estudio sobre ecología de las religiones. Akal, Madrid
36. MacIntyre A (2010) Historia de la Ética. Paidós, Barcelona
37. Macpherson C (2005) The political theory of possessive individualism. From Hobbes to Locke. Trotta, Madrid
38. Martínez Veiga U (1978) An ecological interpretation of religion. In: Antropología ecológica. Adara, La Coruña, pp 203–231
39. Martínez E et al (2015) Neither colonialists nor sympathizers: a response to Eduardo Gudynas. In: https://lalineadefuego.info/2015/10/13/ni-colonialistas-ni-simpaticos-una-respuesta-a-eduardo-gudynas/. Accessed 13 June 2017
40. Næss A (1973) The shallow and the deep, long-range ecology movement. Inquiry 16:95–100
41. Næss A (1987) Self-realization: an ecological approach to being in the world. Trumpeter 4(3):35–41
42. Norton BG (1984) Environmental ethics and weak anthropocentrism. Environ Ethics 6
43. Polanyi K (1989) La gran transformación. La Piqueta, Madrid

44. Riechmann J (2015) Dimensions of religion: common origin of religion and philosophy? http://myslide.es/documents/dimensiones-de-la-religion-jorge-riechmann-origen-comun-de-la-religion-y-la-filosofia-leszek-kolakowski-ha-sugerido-que-tanto-religion-como-filosofia.html. Accessed 7 June 2017
45. Regan T (1981) The nature and possibility of an environmental ethic. Environmental ethics, 3rd edn. University of California Press, Berkeley, CA
46. Reyes Mate M (2011) Tratado de la injusticia. Anthropos, Barcelona
47. Reynolds R, Tanner R (1995) Ecology, meaning and religion. Oxford University Press, Oxford
48. Rollin B (1981) Animal rights and human morality. Prometheus Books, Buffalo
49. Sedláček T (2014) Economics of good and evil. FCE, Mexico
50. Seoane J (2012) Neoliberalism and extractivist offensive. Actuality of accumulation by dispossession, challenges of our America. Rev Theomai 26
51. Sheliepin LA (2010) Far from equilibrium: synergetics, self-organization and catastrophe theory. Popular Sci Ser Phys 5
52. Shiva V (1991) Embracing life. Mujer, ecología y supervivencia (Transl. A. E. Guyer and B. Sosa Martínez). Third World Institute, Montevideo
53. Singer P (1975) Animal liberation. A new ethics for our treatment of animal. Avon, New York
54. Singer P (1995) Practical ethics. Cambridge University Press, Melbourne
55. Stauffer RC (1957) Haeckel, Darwin, and ecology. Q Rev Biol 32(2):138–144
56. Steward J (1955) Theory of cultural change: the methodology of multilinear evolution. Urbana, University of Illinois Press, Illinois
57. Steward J (1977) Evolution and Ecology. Essay on Social Transformation. Urbana, University of Illinois Press, Illinois
58. Svampa M (2011) Minería y neoextractivismo latinoamericano. http://huerquenweb.les.wordpress.com/2011/08/minerc3ada_y_neoextractivismo_latinoamericano-svampa. pdf. Accessed 3 June 2017
59. Svampa M (2011) Neo-developmentalist extractivism and social movements: an eco-territorial turn towards new alternatives? In: Grupo Permanente de Trabajo sobre Alternativas al Desarrollo (eds) Más allá del desarrollo. Abya Yala/Fundación Rosa Luxemburg, Quito, pp 185–218
60. Taylor W (1986) Respect for nature: a theory of environmental ethics. Princeton University Press, Princeton
61. Tawney RH (1972) The acquisitive society. Alianza, Madrid
62. Thompson EP (1995) Habits in common. Crítica, Grijalbo Mondadori, Barcelona
63. van Schaik C, Michel K (2016) The good book of human nature. Basic Books, New York
64. Villarroel R (2006) Nature as text. Hermeneutics and environmental crisis. Universitaria, Santiago
65. Warren K (1996) The power and the promise of the ecological feminism. In: Warren K (ed) Ecological feminist philosophies. Indiana University Press, Bloomington and Indianapolis
66. Weber M (2003) Politics as a vocation. In: El político y el científico. Alianza, Madrid

Economic Ethics

4

Jesús Conill

Abstract

In the current techno-scientific and economic context, characterized by accelerated processes of globalization, it is still difficult to see how economics and ethics are related. Apart from the misgivings and suspicions that still persist, ethics is seen as playing at best a cosmetic role in economic life, to create a good image or reputation. In the face of this widespread view that sees the economic and ethical worlds as incompatible, this chapter recalls some historical milestones that help us better understand our current situation and discover the irreducible ethical background of the economy.

Keywords

Ethics • Economics • Neuroeconomics • Globalization • Global economic ethics

1 Introduction

In the current techno-scientific and economic context, characterized by accelerated processes of globalization, it is still difficult to see how economics and ethics are linked. Most commonly, suspicions and skepticism are aroused, both on the part

This study is part of the research project "Ethics and Democracy in the face of the challenges of Artificial Intelligence" PID2019-109078RB-C22 funded by MCIN/AEI/https://doi.org/10.13039/501100011033 10.13039/501100011033 and the Prometeo 2022 Program for Research Groups of Excellence, CIPROM/2021/072, of the Generalitat Valenciana.

J. Conill (✉)
Universidad de Valencia, Valencia, Spain
e-mail: jesus.conill@uv.es

A. Cortina and M. Correa Casanova (eds.), *Applied Ethics from Medicine to Humor*, Integrated Science 21, https://doi.org/10.1007/978-3-031-59009-2_4

of professionals and citizens. At best, ethics is thought to play a cosmetic role in economic life, in order to create a good image or reputation. In the face of this immediate impression (it has now become a fairly widespread prejudice that the economic and ethical worlds are irremediably incompatible), we should remember some historical milestones which help us to better understand our situation, and to discover the ethical background of the economy [5].

2 Ethical Economics in Aristotle and Adam Smith

The double origin of Economics as a discipline is often alluded to in the fourth century B.C., both in Aristotelian practical philosophy (in Book I of the *Politics*, the "*peri oikonomia*") and in Kautilia's *Arthasastra*.[1,2]

In Aristotelian philosophy, economics was primarily the art of managing the household (the *oikos*), and was part of practical philosophy along with ethics, politics, and even rhetoric. In this conception, economics is one of the human activities, namely that which deals with the administration of goods in order to satisfy needs. Thus, there is rational, natural ("phýsis"), and communal ("koinonia") order involving the good of the domestic community ("oikological") and, by extension, of the city ("political ethics"). From economics (which deals with the use/administration of goods) we must distinguish chrematistics, which is the productive art or the art of acquiring wealth and resources. There are, however, two kinds of acquisitive art or chrematistics: (1) that which procures the resources necessary for life (the wealth that is sufficient to live well, but which is not unlimited); and (2) that which seems to have no limit. This second kind of chrematism goes against the natural rational order because it has no limit. Hence the traditional rejection of usury.

Thus, on the ancient classical horizon, in the Aristotelian version, economics is part of practical philosophy in intrinsic connection with ethics and politics. It has a communitarian character and a holistic approach, in which the whole is prior to the parts; therefore, one can really speak of a common good, of a common interest, which holds the parts together and is at the service of the satisfaction of needs, which are by nature limited. This is one of the crucial differences with the modern approach to the economy, in which it is not possible to affirm in the same way that needs are by nature limited, because modernity tends to confuse them with desires, which can be unlimited (unlike vital needs, which are limited); nor will it be possible to maintain the holistic and communitarian sense. Despite this, the link between economics and ethics and politics will continue to be maintained within a new practical philosophy.

---

[1] It would consist of "instructions on material prosperity" and a certain "science of wealth."
[2] He was advisor and minister to Chandragupta, emperor of India (fourth century B.C.).

Indeed, in classical modern economics there remains a link between the moral and technical sense of economics, as shown by the paradigmatic case of Adam Smith (1723–1790),[3] professor of Moral Philosophy at the University of Glasgow (Scotland) [from 1752]. In 1759 Smith published his book *Theory of Moral Sentiments* (in which his ethical theory is based on the "sympathy" between human beings), and in 1776 a book considered foundational of modern economics: *Inquiry into the Nature and Causes of the Wealth of Nations.* Thus, the emergence of classical modern economics is also linked to moral philosophy and will be understood as "political economy," causing the historical context of economics and the meaning of ethics to change.

It is the epoch in which the mercantile economy emerges with force, based on exchange, due to the increase of commerce: the "sweet commerce," because it is better to trade than to war. Also in the economy, the Kantian appreciation of politics and law would be applied, showing that preferring trade to war would be understood "even by a people of demons, as long as they had intelligence" [9]. Given the current international panorama, we have not learned even this minimum yet!

A system of economic organization emerges, which is not separated from society, but rather configures society as commercial and mercantile. In such a way, economic activity is intermingled with the moral elements of society, whose new starting point is not the whole, but the individual as a subject (moral, social and economic, interested in his own), and a first modern way of understanding freedom as rooted in understanding oneself as an individual.

In this new modern context, relationships between individuals are established mainly through the contract, which occurs in politics (as is evident in the various theories of the contract as a legitimation of political life) and in the economy. Exchange in commercial society is configured by means of contracts in which mutual advantages are sought. Therefore, social and economic order cannot be separated because, ultimately, it is about free individuals within a political and social framework; hence, the modern classical economy is understood as a political economy, based on a modern ethic that starts from the individual exercising his or her freedom.

That is why, when Adam Smith's famous passage "it is not the benevolence of the butcher, the brewer, or the baker that procures our food, but the consideration of their self-interest.[4] We do not invoke their humanitarian feelings…; nor do we speak to them of our needs, but of their advantages" ([50]: 17) is quoted, it is necessary to contextualize it and complete this perspective with others that also appear in the same work. In the exchange, the deciding factor is not the need, but the mutual advantage.

---

3 And the relevant interpretations of the emergence of modern economics by Weber [57], Sombart [52], Tawney [54], Robertson [41], Chafuen [4], Koslowski [29], etc., have also stood out.

4 We must distinguish "self-love" from "selfishness" (selfishness, incompatible with sympathy).

Smith describes in a personified way the mechanism of exchange (within a chapter entitled "On the principle that motivates the division of labor") as an expression of a "propensity to exchange, barter and negotiate one thing for another," a propensity of which he wonders if it is not probably "the consequence of discursive faculties and language" ([50]: 16). Thus,

> in a civilized society [one] needs cooperation and assistance at every moment .... In most circumstances man demands the help of his fellow men, and in vain can he expect it only from their benevolence. He will obtain it with greater certainty if he interests [others] in his favor ... and makes them see that it is advantageous for them to do what he asks of them. He who proposes a deal to another is making him one of those propositions. *Give me what I need and you will get what you want* is the meaning of any kind of offer, and so we get from others most of the services we need. ([50]: 16–17)

In this order of exchange, the motivation is certain cooperation.

But in this economic framework of exchange, although "no one generally sets out to promote the public interest, nor does he know to what extent he promotes it," when security and self-interest are sought, one

> is led by an invisible hand to promote an end which did not enter into his intentions. But it is no harm to society that such an end does not enter into his purposes, for in pursuing his own interest, he promotes that of society more effectively than if it entered into his designs. There are not many good things that we see executed by those who presume to serve only the public interest. ([50]: 402)

Two important issues arise from this in subsequent interpretations: (1) the question of unintended consequences in the shaping of social relations; and, (2) the presupposition of a basic harmony in society, or conversely, that of an irresolvable conflict.

In any case, an all too common error has been to confuse Smith's position with that of Mandeville; we must not confuse self-interest and "the tendency to exchange" with a vice. Smith himself makes this clear in part VII of *The Theory of Moral Sentiments* ([51]: 534 ff.) on "licentious systems." For Smith, Mandeville's system eliminates the distinction between vice and virtue, so this tendency is "wholly pernicious": "the great fallacy of Dr. Mandeville's book [*The Fable of the Bees*] is that he represents any passion as fully vicious... and by means of this sophistry he establishes his favorite conclusion: private vices are public benefits" ([51]: 543). Adam Smith rejects Mandeville's position, because he considers that it is a nefarious interpretation of human motivation, since it does not distinguish between different types of passion (the just and reasonable for the noblest; the unjust, absurd, and ridiculous for the lowest and the frivolous—such as vanity) ([51]: 538 ff.).

On the other hand, it is necessary to complete Smith's vision on the value of the market mechanism with the need to submit to a critical study the role of markets ([47]: 156 ss.). Smith gives a severe warning on the use of money by "prodigals and adventurers" because they provoke and foment social waste. His argument, as interpreted by Sen, is that market signals can be misleading, and the consequences of the free market can mean a great waste of capital. Smith was concerned about the problem of social waste and the loss of productive capital, and that the pursuit of private profit, driven by short-sighted motivations, would lead to social losses. He pointed out that in some circumstances the pursuit of private profit runs counter to social interests. This concern is still relevant today: it warns us of the limits of the market mechanism and of the need to maintain a critical attitude of discernment. Controls are needed to make the "system of freedom" in the "commercial society" feasible.

3 Economy Without Ethics?

The modern process of rationalization, demystification, and disenchantment has affected the way economic activity is conceived and practiced. As the effective process of modern rationalization progressed, the positivist and technical side of economics became predominant. It lost sight of the ethical sense to such an extent that the conviction has become widespread that economic rationality (which together with technological rationality has become the model of rationality for many) excludes any ethical approach other than the utilitarian one. It has caused a split between the ethical and the techno-scientific economics, which constitutes a social problem. Several factors have intervened in this modern process, among which the axiological, the epistemological, and the professional-institutional ones are worth mentioning.

First, the economy was liberated from the tutelage of morality (and religion) in the context of traditional societies. Modernization (capitalist or socialist) produces this autonomization of the economic sphere. Together with technological innovation, this has been perhaps the most powerful driving force in replacing the traditional symbolic universe with another way of understanding the world and organizing it.

One of the consequences has been the axiological change. For example, in one of its versions, the "profit motive" began to be perceived as "natural" and morally justified; moreover, it was progressively becoming something beneficial and advantageous for human interrelation and for the functioning not only of the economy but also of society. It even changed the model of the human being in the collective imagination: people are understood as individuals, rational egoists, calculators of advantages and benefits, and apparently insensitive to moral arguments.

This process of rationalization (demystification and disenchantment) has generally been interpreted as deaxiologization under the predominance of technocratic positivism, which defends "axiological neutrality" (*Wertfreiheit*) and favors the loss of ethical sense in economic activities. However, it is also possible to interpret this process as a transvaluation, not precisely in the style of the Nietzschean "transvaluation" (*Umwertung*), but of a very different type, because new values emerge that govern real life and create modern economic culture, in which the economicist transvaluation has won, in different versions, in addition to the technological one. Not recognizing this axiological load turns that attitude into a deforming ideology of reality [1, 22, 48].

Second, together with the axiological change in the social mentality, there is an epistemological change in the conception of economics as a science. The model of interpretation of economic activity tends to be mechanistic (under the influence of the reigning science of modernity, which is mathematical physics): economic agents, who seek their own profit, are considered "forces" within an event understood as a natural process, a "field" of forces, from whose dynamism it is possible to establish "laws."

With the above, the economy ceases to belong to practical philosophy, it ceases to be a social and historical science, and to submit primarily to a process of formalization and mathematization within the mechanistic approach reigning and prevailing in modern science. Consequently, it is no longer possible to apply ethical categories, such as social justice, to economic mechanisms thus understood. A paradigmatic example is Hayek's intelligent position in this respect, according to which there is no need for ethics in the natural occurrence of modern economy [23, 24]. The economic system is conceived as a functional mechanism, in which the presumed "homo oeconomicus" functions as a reactive machine to the stimuli of the economic field (of *inputs* and *outputs*). It is the triumph of the algorithmic machine in the corresponding field of forces.

Third, this axiological and epistemological change is translated into practical life and transferred into social life through the training of professionals, the design of organizations and the configuration of institutions.

As a consequence of this modern process, there has been a growing tendency towards the economization of all areas of life: the human being as an economic agent, understood under the figure of homo oeconomicus (it would be better to say "homo chrematisticus"), interprets reality through the lens of "cost–benefit." He starts from the fact that resources are scarce, then tends to satisfy his preferences trying to maximize the benefit through the most efficient strategies. This is the approach that has become widespread in public and private life. In this way, there has been a growing process of absorption of phenomena that were not strictly economic (i.e., market) by economic rationality—by the economic analytical perspective—applicable to any field of human behavior. This is so, since every action and omission (in whatever field) can be considered from the perspective of opportunity costs and cost–benefit analysis. Economics provides a general theory of rationality and analysis of human behavior; in short, a framework of intelligibility to explain (presumably) all factors of human life. This is what has been called "economic imperialism," and has even attempted to provide an economic theory of morality (as has been done for politics, culture, education, and personal life).

On the other hand, the conviction seems to have been reached that there is no encompassing ethics capable of rationally grounding universally binding moral norms, thus calling into question the pretensions of ethical cognitivism (of Kant,

Rawls, Apel and Habermas). For example, instead of a foundation of morality based on a strong concept of "reason," Homann, tries to found moral binding on the basis of "interests" and motivations. And, according to Luhmann, ethics is irrelevant for the communicative processes that are established in functional systems [25, 31].

3.1 New Forms of Economic Ethics

For some time now we have been witnessing a critical review of economic theory and new proposals beyond conventional economics. There is a recognition of the current crisis of the economy and more and more critical voices are being heard.[5]

There are those who insist that Economics is not a physical science, but a social and historical science—that is, a human science. Some even go so far as to say that it is not a predictive science, but a reconstructive and indicative one. Hence, it is necessary to reflect on the assumptions on which it operates in each of its formulations. For example, the presupposition of "rational behavior," which has allegedly served to predict behavior, is especially criticized ([46]: 29).

There are proposals of all kinds: from those who propose a change of the economic system, to those who advocate a change of paradigm or approach in Economics. They consider that conventional economics is not useful to solve the new problems because it has been impoverished, having lost its sense of political economy, and especially, because it has been separated from ethics. A new economic ethic is needed.

Hence, a number of initiatives have emerged that have attempted to explicitly incorporate ethics by systematically linking economics and ethics. For example, social liberalism (from Rawls onwards); the ethics of constitutional economics (J. Buchanan); the new economic institutionalism (North); communitarian socioeconomics (Etzioni), economic republicanism (Sandel, Ackerman, Pettit); "social capital" (Putnam); the economic theory of morality (Homann); economic and business ethics based on dialogic praxis (Steinmann); Ulrich's communicative (integrative) economic ethics [5, 37]. More recently, other initiatives have emerged: the psycho-economics of Kahneman and Tversky [27]; and the new theory of the commons of Ostrom [40].

Even more significantly, we must highlight the contribution of Amartya Sen's "capabilities approach" [16]. This new economic approach has an ethical background beyond utilitarianism (in any of its forms). Sen's economic thought is an indication that the field is opening up for new forms of economic ethics: this author explicitly maintains that "the nature of modern economics has been substantially impoverished by the estrangement that exists between economics and ethics" ([46]: 25).

[5] See, for example, A. Hirschmann, P. Ormerod, L. Thurow, D. Schweickart, L. de Sebastián, R. Velasco, E. Ostrom, S. Zamagni, D. Rodrik, J. Sachs, Th. Piketty, J. Tirole, etc.

Another good example (which can also be related to Sen's influence) are the "Human Development Reports" of the UNDP (United Nations Development Programme), where both the fundamental approach and the poverty indicators are qualitatively different from those of conventional economics. For example, the new human development indicators focus on the struggles for human freedoms.[6]

The basic issue of Sen's new approach is to make us aware that when we talk about wealth and prosperity (and poverty), it is not enough to talk about money or GDP per capita. We must also consider other determinants and vital factors; we need to have other sources of information, for example, about the quality of life and about what capacity people have to lead their own lives. What is their real level of freedom. Since this is only possible within a certain framework of justice, it is necessary to have information about security, life expectancy, health, medical services, education, work, freedoms, family relations, the situation of women, and so on. "Life is more than a set of commercial relations," and "the human being… is an 'unfathomable mystery,' which cannot be fully expressed in a 'tabular form'" ([35]: 16). It requires information on "what people can do and be."

Undoubtedly, it is more complex to measure and evaluate these dimensions, but it is necessary if we want to "know how people are doing" in their lives in different parts of the world. Hence the need for new indicators with which to measure the quality of life. It is no longer sufficient or convenient to continue with a utilitarian approach; it is necessary to search innovatively for other instruments to attend to the diverse activities that make human life flourish. This is one of the purposes of Sen's *approach of the capacities*: "to find adequate criteria to value the quality of life" ([35]: 17). The utilitarianism behind the theory of revealed preferences is not enough, neither are its axioms and algorithms.

However, the most significant basis of this capabilities approach (even if it is within the most relevant traditions of economics and ethics) is established by "the perspective of freedom," which reintroduced an unavoidable ethical background in economic theory: hence it is appropriate to call it "ethical economics," whose ingredients are freedom and quality of life, freedom and justice, without which real freedom is not possible. We must not confuse *welfare* and *well-being*, that is, *well*-being and *good*-being; the latter is the one that moves away from the utilitarian conception of well-being and relates to the condition of the person (referring to other aspects, such as capabilities, opportunities, advantages, hence the term "quality of life" is used).

[6] See, for example, *Human Development Report 2000*, chap. 2; also Goulet [21], Martínez Navarro [32] and Pedrajas and Choritz [36].

4 Expanding Economic Ethics from the Neurosciences

Economics is also being affected by the current neuroscience revolution, so that important research in "neuroeconomics" and "neuromarketing" has already been developed (in a similar way to what has happened in other disciplines, due to the influence of biotechnological innovations in neurology, which allow the detection and visualization (imaging) of brain activity, not only through electroencephalograms, but also with the new techniques of PET[7] and fMRI[8]) [11, 12].

Despite some initial reluctance, the growing neurological innovation of recent decades is already bearing fruit, making possible the objective study of subjective phenomena, specifically brain activity in decision-making, i.e., the neurobiological processes underlying decision-making ([53]: 34, [18]). One of the consequences of these studies is that they also contribute to revise the standard concept of economic rationality, which is focused on utility maximization. The study of the neurobiological conditions of choice is changing the algorithmic processes that attempted to explain decision-making. In this way, the neuroeconomic approach could serve as an alternative to the rule of normative and axiomatized theories of contemporary economics. For one thing, these new neurological studies play with Kahneman and Tversky's *prospect theory*,[9] which is a non-normative form of *expected utility theory*. They also aid in bringing the economic theory of the so-called "econos" (rational agents) closer to the real world of "human" agents.

The contributions of the neuroeconomic approach to the study of uncertainty, risk assessment, aversion to ambiguity and loss, biases, the relevance of feelings, emotions and values in decisions, etc., are already many. The investigation of the brain bases of the presumed rational agent (the neurobiological processes underlying decision-making) offers a new way to overcome motivational homogeneity (which, for example, Sen already overcame with his approach to capabilities in relation to moral motivation).

This broadening of motivations has been dramatically applied to investment activity. By studying brain activity in financial investments, some research has shown that our investment brains often lead us to do things that don't make logical sense, but make emotional sense. It is not that we are irrational, rather that calculation is combined with reactions to risks, rewards, memories, and hunches that make us feel hope, greed, surprise, fear [61], panic, regret, happiness, and so on.

Consequently, through all these new studies, standardized economic rationality has been revised, establishing its limits and its necessary flexibility. It is even possible to ask whether we should not speak of a "neurorationality" of choice, with important consequences for some ethical aspects, such as: (1) we are not

---

7 Positron emission tomography scanner, which measures blood flow in the brain.

8 Functional MRI, which tracks blood flow in the brain based on blood oxygenation (the "bold signal").

9 *Prospect theory* as translated in Kahneman [27].

optimal rational decision-makers who maximize utility; (2) we need to establish relationships of trust, which can be favored by some neurotransmitters, such as oxytocin; (3) there are neurological bases for cooperation and even for the sense of justice in economic activity.

Some neurobiological experiments have provided data showing a connection between the medial prefrontal cortex and some cooperative strategies that deviate from standard normative prediction, which has led to the hypothesis that the non-normative model of cooperation may not be based in prefrontal cortex circuitry.

Likewise, the neuroscientific study of *reward* in decisions offers data (although a coherent theory for its overall interpretation is lacking) that call into question the interpretation of individuals as optimal decision-makers who maximize utility. A very relevant aspect is the role of neurotransmitters in the reward process, for example, dopamine. This is an issue of enormous importance today in the growing number of studies on the relationship between the brain and decision-making. For what is represented in the brain is neither pure information nor pure utility (or preference), but a fusion of information and valuation (i.e., not exclusively "utility"). It is as if certain neurons in the orbitofrontal cortex say "this is such a thing, and I want it (or I like it)," thus determining "the *reward value* of the recognized object," which in turn depends on multiple factors (personal life and circumstances).

Indeed, there are substances that alter behavior with respect to the canons of normative economic theory. This is the case with oxytocin, which increases trusting behavior [60]. According to Zak, it plays a decisive role in economics, insofar as "interpersonal trust is the most important factor that economists have discovered to explain why countries are rich or poor," since trust favors prosperity. Fehr and his team discovered the effects of oxytocin on behavior and social interaction; they showed that high levels of this hormone in the blood improves the ability of individuals to trust others. Oxytocin functions as a natural lubricant in social life, helping overcome the fear of being deceived. But some studies have also highlighted that the administration of oxytocin alters the behavior of the subjects, but does not affect the will.

Along with studies on the neurological basis of cooperation and trust, we also find research on the neural basis of the sense of justice applied to the field of economic decisions. The famous experiment called the *ultimatum game* [34] stands out. Alan G. Sanfey made a study based on this game, which consisted of putting the players in pairs [45]. The first was given an amount of money, leaving it up to him to choose what he wanted to give to the second. The second, who knew the total amount, could accept the offer or not. If he accepted, they both earned what they had agreed, if he rejected the offer as unfair, they would both lose the money. The rational choice (according to the canons of standard maximization for the second player was to accept the offer made to him, since something is better than nothing. However, the sense of fairness took precedence over rational maximization. The second players preferred to get no benefit rather than accept an offer they considered unfair. This experiment shows that human beings are not rational decision-makers who are governed by maximization; in the final resolution, other more important factors are involved in decisions ([13]: ch. 4).

The contribution of the neuroeconomic approach consists of studying how the brain intervenes in decision-making. An activation of the cerebral hemispheres has been detected when subjects decide about the acceptance or rejection of unfair offers in negotiation. Moreover, from a neuroeconomic point of view it seems that correlations can be established by dissociating judgment and choice, because choice implies that the decision maker has to assume the costs and benefits, whereas mere judgment is not yet associated with the assumption of costs and benefits.[10] The scope of ethical issues is thus broadened through the innovation of neuroeconomics.

5 Global Economic Ethics

Economic ethics must combine two levels of reflection: the theoretical level, both in economic and ethical perspectives, and the level of the globalizing processes as they are actually taking place. However, the horizon of globalization does not consist of mere factuality, but is also full of possibilities, opportunities, and ideals (such as freedom and justice). For this reason, it is necessary to orient the processes in a humanizing sense, making ethical globalization truly possible.

But is it possible to speak of freedom and justice in the economic processes that have already been set in motion, or are they intrinsically bad? We must go beyond the Parmenidean approach (which only distinguishes between "being" and "not being"), and adopt an attitude like the Aristotelian one, more in line with ethics, which tries to detect what "can be" (the real possibilities), that is, the dynamisms that can be oriented to improve human life.

One of the most decisive aspects of globalization has been the financialization of the economy. Therefore, the question is whether it really promotes the moral values of freedom and justice, or whether it favors other objectives to the detriment of these values, which are basic for quality human coexistence. For example, we must realize that if financialization promotes short-termism in capital investments it will have pathological effects, because it deforms economies, produces instability, uncertainty, concentration of power, growing vulnerability (especially of the weakest), unjust inequalities, poverty, and exclusion, which already begin by neglecting human development.

In this context, the proposals for some taxes that would serve to control capital flows, and that favor transparency have been very significant (such as the famous Tobin tax). Additionally, given the enormous amount of capital that moves in the international financial markets in the short term, these taxes could even to raise funds to be used to alleviate the pathological effects of globalization itself, and to invest in the human development of the least favored.

[10] This distinction reminds us of the Aristotelian distinction between cognition and choice (cf. *On the Soul*, II, 11: the intellectual faculty alone does not produce movement), for cognition by itself cannot produce action, since in order to influence behavior the cognitive system has to operate through the affective system.

It is true that many are fascinated by profit growth due to new opportunities in the fields of production and marketing. These opportunities are the result of technological innovation, changes in transportation and communications, and the free circulation of capital. In the new financial markets, in which astronomical figures move every day, practically without interruption. We should remember, however, that all of this financial growth has produced a persistent climate of global instability, and lack of control; we should remember that profits have not been channeled towards the real economy of goods and services, nor towards the creation of employment, nor towards the development of the most disadvantaged and impoverished among us, but primarily and persistently towards financial speculation. Hence, the process of globalization has been accompanied by increased inequalities between the rich (powerful) and the poor (*aporoi*), and also by a growing *aporophobia* (aversion to the poor) in our "advanced" societies [14].

Is the economy marked by the financial attitude that is taking us back to the most perverse tendencies of capitalism? Certainly the transformation of capitalism has been very deep; but it could be that new contradictions are emerging from the development of the financial system and other forms of domination of people and peoples, coming mainly from the growing weight of the globalized financial sphere. Therefore, it is not idle to ask whether financial capitalism can be a social, inclusive, and ethical capitalism, or whether it is impossible by its very nature to tame it again, and turn it into an instrument at the service of a just and decent society, of a good society.

Is the financial system a system lacking in morality because of its own structure and ways of functioning? Is it destined to increase the number of "evils" of which capitalism has been accused, since it does not place the economy at the service of people but at the service of the profit motive and (even more crudely) of unlimited and rapid profit in the shortest possible time, without further considerations? Is it the most appropriate field for the promotion of the *homo oeconomicus* (chrematistic), under the consideration that it is the best expression of human nature (a calculating and utilitarian being)? Does this further reinforce the "culture" of inequality, of competition, of struggle, and the instrumentalization and commodification of communication? A serious danger of the new international situation is that capitalism, which for decades tried to appear as having a human face, now believes it does not need one and presents itself with traces of growing inhumanity.

There are those who insist that as the processes of globalization are taking place, they foster selfishness and greed (since they are based on the pursuit of profit), and throw a large part of society into poverty and unemployment, thus creating or maintaining profound inequalities; however, there are also those who highlighted, among the virtues of globalization, its effectiveness in producing wealth and well-being. Others even point out that the economy thus develops a peculiar and typically modern moral background: that of interest. This has become an essential element of modern social ethics, since for many it constitutes an effective guarantee of social order. In this sense, the desire to improve one's own situation constitutes an inexhaustible source of benefits for society, since it encourages creation, innovation and risk-taking.

In the context of the current transformation of capitalism, the debate between contemporary ethics in relation to the economy cannot avoid asking if it has no more room for social and economic justice, if this is totally absorbed by the focus on interest and mutual advantages (which leads irremediably to the mercantilization of justice), or if it is possible and reasonable to continue trying to incorporate justice to the alleged "system of freedom," so that it has a glimmer of becoming real (effective) in people's lives ([5]: 224–237).

6 New Perspectives and Initiatives for Economic Ethics

(1) *Awareness of the need for universal ethics.* For some time now, voices have been calling for an ethical perspective to give proper direction to the processes of globalization, and to face the new problems of the global order. It is a question of a universal ethics, capable of providing an axiological and normative framework for international politics and economics, and for the immense potential of new information and communication technologies. Hence, the "need for a universal ethics as a global macroethics of co-responsibility in the era of science and technology" ([1]: 341–413, [2]: 91–218; [6, 7, 10, 26]), and of cosmopolitan character [15] has been increasingly highlighted.

There exists a need for an ethics of responsibility, or better, of co-responsibility guided by justice, care, and solidarity: not only personal, but social and institutional; not of disinterest, but capable of sifting and orienting the complicated world of interests; not only based on convictions and principles, but attentive to the consequences, dangers, and risks of actions and decisions in the exercise of the growing intervening power in nature and in society.

(2) *Overcome the financialization of the global economy* in order to bring about a new, more humane and humanizing way of life, centered on the real needs of people and the real problems of humanity.

The exponential growth of financial markets has had positive consequences, such as the high availability of credit to finance consumption and investment, growth and employment. However, imprudence and inefficient supervision has led to fatal risk-taking, resulting in instability. Financial excessiveness requires correction and rational control of activities [17, 19]. As Samuelson himself stated, "unregulated market systems end up destroying themselves" and are "the primary source of our problems today" [44]; financial capitalism has run amok. We need to think again and follow Roosevelt's advice: "I would rather bail out those who produce food than those who produce misery." Recently, even those who contributed to the current financialization, such as former Citibank CEO Sandy Weill, have acknowledged that "allowing commercial banks to merge with investment banks was a mistake." Therefore, the real economy must be rebuilt and strengthened, and "those who want to take risks should do so on Wall Street" and with their own money, not with all the money of all their depositors (including those who do not want to participate in this dangerous game.

Financial transaction taxes, such as the Tobin tax [55, 56], can help to discourage transactions that favor the flow of speculative capital in very short-term transactions, and deter speculators; they can thus serve the objective of limiting the excessive power of financial markets. Tobin himself made it clear that he was influenced by Keynes, who in Chapter XII of his *General Theory of Employment, Interest and Money* had already devised a tax on financial transactions, because excessive speculation increased market volatility. Financial markets needed to be better regulated so that the basic requirement of justice could be met, starting with the satisfaction of needs.

(3) *The management of globalization* (governance?) [20]. The processes are the result of international decisions and agreements. Therefore, we should pay attention to the consequences. They are the result of institutional decisions for which there are individuals who are responsible. It is necessary to reflect on the way of life promoted by these world economy transformation processes: whether or not this way of life favors real freedom for all; whether it favors justice, or if generates more unjust inequalities; whether it contributes to creating a climate of solidarity, or favors economic war (and not only economic) and the global jungle.

Reputed scholars, such as Castells, recognize the close linkage between economics and politics in the current phase of development. There is a complex integration between Politics and Economics, given that the economy (global and international) is "a highly politicized economy" ([3]: 117). Increased market competition on a global scale takes place "under conditions of managed trade" ([3]: 117). In that context, business innovation is combined with governance strategies. The new economy, based on socio-economic restructuring and technological revolution, will be shaped by state-driven political processes.

A new economic role of the state in an era of deregulation is thus emerging. Although some speak only of the weakening of the state (or even its disappearance), we need to reflect more deeply on its new role and new functions—on how it will transform itself into an efficient institution, suited to the current development of the global society. What is crucial in the new information society, according to Castells, is the complex integration of political institutions and economic agents (the latter increasingly globalized). Hence the need for institutional synergy in an environment and times accustomed to conflict, but without confusing it with the connivance between economic, political and media power, since in the financial and informational development of the economy all powers are closely linked.

(4) The trend towards economic inequality, which has serious implications for democratic political life, must be addressed. Thomas Piketty's book, *Capital in the twenty-first century* highlighted that the cause of inequality is the way in which capitalism has developed [38]: the rate of capital profit is systematically higher than the rate of growth of the economy, which is what benefits the majority of people. Whether or not his etiological diagnosis is right, Piketty's contribution highlighted the need to attend to this aspect of distributive justice. Its resolution requires a new way of managing economic activities within an ethical horizon of global justice, correcting its perverse tendencies (e.g., the excessive financialization), and redesigning the institutions of the "Social State of Justice," which should not be confused with that of "welfare" ([8]: chap. III).

(5) Along the same lines of rectifying abuses in the area of finance, the managing director of the International Monetary Fund, Christine Lagarde, has accused the banks of slowing down and attempting to halt international reforms [30]. In her speech at the "Conference on Inclusive Capitalism," she denounced the slowness with which the reform of the global financial system is being implemented. Trust in the financial system has been lost because it is producing increased inequality. Apparently, while the 85 richest people in the world could fit on a London bus, they control as much wealth as the poorest half of humanity. There should be "more inclusiveness in economic growth and more honesty in the financial system." These terms (social inclusion and honesty) imply an ethical horizon for the economy. And one of the priority proposals that Lagarde has defended is to "put an end to *too big to fail*," because it goes against the sense of justice.

(6) "Seeds" of fair economy: new initiatives of economic ethics. Initiatives are emerging from various sectors of civil society [33], although not always sufficiently addressed by the mainstream of academic socio-economic analysis. They are proposals for business, consumption, and investment models, in which economic activity plays an instrumental role in solving specific societal problems to meet basic social needs, because they aspire to a fairer society. They do not prioritize profitability, but aim to achieve an economy at the service of people. They try to recover the social sense of financial activity and of money itself, supporting projects of special social relevance (in favor of the disadvantaged or those excluded from the conventional financial sector), or creating ethical financial institutions (ethical banking). In all of them, the aim is to show that it is possible to put into practice alternatives for a fair economy for another way of life and society [28, 49]. They are proposals that nourish hope.

(7) Some of them are becoming institutionalized, such as those promoting "Corporate Social Responsibility" [39, 59], beyond Business or Corporate Social Responsibility, as there are those who consider that "the role of business in society is the most important public policy issue of the twenty-first century. The company is increasingly shaping social values and norms...." ([58]: 1–2).

The framework of this proposal is that of a "civil economy" as a social innovation, whose roots are found in the civic humanism of the fifteenth century, and its development in the Italian Enlightenment of the eighteenth century. Here, the economy is not identified with the market, but is sustained by values outside the market,and politics is not identified with the state, but is likewise sustained by values outside the state.

(8) Another attempt at international institutionalization of the new ethical impulse of the economy was achieved in 2011, when the United Nations Human Rights Council endorsed the "Guiding Principles on Business and Human Rights," explained by John Ruggie, the Special Representative of the UN Secretary-General for Business and Human Rights. Its main guidelines are to "protect, respect, and remedy": protect from abuses, respect human rights and remedy (redress) victims [42, 43]. This framework is a big step, although still insufficient,hence the call for a binding international convention to give effect to the Guiding Principles through some regulation of the activities of transnational corporations in favor of human rights.

Some believe that economic convergence is the driving force behind cultural convergence through globalization, which is producing both economic and cultural homogenization. If true, we must ask whether what is being imposed is the commodification of the whole of life, eroding pluralism, and causing an impoverishment of forms of existence, or whether we choose to promote an authentic economic ethic, based on global justice and human rights.

References

1. Apel K-O (1985) El a priori de la comunidad de comunicación y los fundamentos de la ética. In: Apel K-O (ed) La Transformación de la Filosofía, vol II. Taurus, Madrid, pp 341–413
2. Apel K-O (2003) Globalización y necesidad de una ética universal. In: Cortina A, García-Marzá D (eds) Razón pública y éticas aplicadas. Tecnos, Madrid, pp 191–218
3. Castells M (1997) La era de la información, vol I. Alianza, Madrid
4. Chafuen A (1991) Economía y ética. Rialp, Madrid
5. Conill J (2004) Horizontes de economía ética. Tecnos, Madrid
6. Cortina A (1985) Razón comunicativa y responsabilidad solidaria. Sígueme, Salamanca
7. Cortina A (1986) Ética mínima. Tecnos, Madrid
8. Cortina A (1997) Ciudadanos del mundo. Alianza, Madrid
9. Cortina A (1998) Hasta un pueblo de demonios. Ética pública y sociedad. Taurus, Madrid
10. Cortina A (2010) Justicia cordial. Trotta, Madrid
11. Cortina A (2011) Neuroética y Neuropolítica. Tecnos, Madrid
12. Cortina A (ed) (2012) Guía Comares de Neurofilosofía práctica. Granada, Comares
13. Cortina A (2013) ¿Para qué sirve realmente la ética? Paidós, Barcelona
14. Cortina A (2022) Aporophobia. Why we reject the poor instead of helping them. Princeton University Press, Princeton, NJ and Oxford
15. Cortina A (2021) Ética cosmopolita. Paidós, Barcelona
16. Cortina A, Conill J, Enderle G, Pereira G, García Marzá D, Crocker DA (2013) Sen's capabilities, poverty and economic welfare. In: Luetge C (ed) Handbook of the philosophical foundations of business ethics, vol 2, Part 9. Springer, pp 659–740

17. Galbraith K (1991) Breve historia de la euforia financiera. Ariel, Barcelona
18. Glimcher PW, Camerer CF, Fehr E, Poldrack RA (eds) (2009) Neuroeconomics: decision making and the brain. Elsevier
19. Gómez-Bezares F (2001) Ethics, economics and finance. Government of La Rioja
20. González E (ed) (2013) Ética y gobernanza. Granada, Comares
21. Goulet D (2006) Development ethics at work. Routledge, London and New York
22. Habermas J (1984) Ciencia y técnica como ideología y Conocimiento e interés. Tecnos, Madrid
23. Hayek F (1982) El espejismo de la justicia social. In: Derecho, legislación y libertad, vol II. Unión editorial, Madrid
24. Hayek F (1990) La fatal arrogancia. Unión editorial, Madrid
25. Homann K (2003) Anreize und Moral. LIT, Münster
26. Jonas H (1995) El principio de responsabilidad. Herder, Barcelona
27. Kahneman D (2012) Pensar rápido, pensar despacio. Debate, Barcelona
28. Kinsley M (2009) Creative capitalism. Simon & Schuster, New York
29. Koslowski P (1986) Ethik des Kapitalismus. Mohr, Tübingen
30. Lagarde C (2014) Economic inclusion and financial integrity-an address to the conference on inclusive capitalism. International Monetary Fund. http://www.w-t-w.org/en/wp-content/uploads/2014/05/IMF_Lagarde_Speech_27.5.14.pdf
31. Luhmann N (1984) Soziale Systeme. Suhrkamp, Frankfurt
32. Martínez Navarro E (2000) Ética para el desarrollo de los pueblos. Trotta, Madrid
33. Moreno JA (2014) Semillas de economía alternativa. Mediterráneo Económico, no 26:291–307
34. Nowak MA et al (2000) Fairness versus reason in the ultimatum game. Science 289:1773–1775
35. Nussbaum MC, Sen A (coords.) (1996) La calidad de vida. F.C.E., Mexico
36. Pedrajas M, Choritz S (2016) Getting to the last Mile in least developed countries. United Nations Development Program, New York
37. Pereira G (2013) Elements of a critical theory of Justice. Palgrave Macmillan, New York
38. Piketty T (2014) El capital en el siglo XXI. F.C.E, Madrid
39. Porter ME, Kramer MR (2011) Creating shared value. How to reinvent capitalism and unleash a wave of innovation and growth. Harv Bus Rev 89(1–2):81–96
40. Ramis A (2017) Bienes comunes y democracia. Crítica al individualismo posesivo. LOM ediciones, Santiago de Chile
41. Robertson HM (1973) Aspects on the rise of economic individualism: a criticism of Max Weber and his school. A.M. Kelly, Clifton
42. Ruggie J (1998) Constructing the world polity: essays on international institutionalization. Routledge, New International Relations
43. Ruggie J (2013) Just business: multinational corporations and human rights. Norton, New York
44. Samuelson P (2008) Adiós al capitalismo de Friedman y Hayek. El País, pp. 26–10
45. Sanfey AG (2003) The neural basis of economic decision-making in the ultimatum game. Science 300:1755–1758
46. Sen A (1989) Sobre ética y economía. Alianza, Madrid
47. Sen A (2000) Desarrollo y libertad. Planeta, Barcelona
48. Scheler M (1973) Sociología del saber. Siglo Veinte, Buenos Aires
49. Shiller R (2012) Las finanzas en una sociedad justa. Ediciones Deusto, Barcelona
50. Smith A (1982) La riqueza de las naciones. F.C.E., Mexico
51. Smith A (1997) La teoría de los sentimientos morales. Alianza, Madrid
52. Sombart W (1979) El burgués. Alianza, Madrid
53. Suhler C, Churchland P (2011) The neurobiological basis of morality. In: Illes J, Sahakian BJ (eds) The Oxford handbook of neuroethics. Oxford University Press, pp 33–58
54. Tawney RH (1954) Religion and the rise of capitalism. American Library, New York
55. Tobin J (1974) The new economics one decade older. Princeton University Press, Princeton, NJ
56. Tobin J (2001) Interview. Der Spiegel, 3 September 2001
57. Weber M (1975) La ética protestante y el espíritu del capitalismo. Península, Barcelona

58. Zadek S (2001) The civil corporation. The new economy of corporate citizenship. Earthscan, London
59. Zamagni S (2014) El reto de la responsabilidad civil de la empresa. Mediterráneo Económico 26:209–225
60. Zak PJ et al (2005) Oxytocin is associated with human trustworthiness. Horm Behav 48:522–527
61. Zweig J (2007) Your money and your brain. Simon and Schuster Paperbacks, New York, London and Toronto

Business Ethics: Towards an Ethical Management of Social Responsibility

5

Domingo García-Marzá

Abstract

Today, business is a clear example of the new protagonism of civil society, of a new distribution of responsibility for the public sphere, as shown by the direct call for its current involvement in the achievement of the Sustainable Development Goals. In other words, business is increasing its capacity to act, and with it its social responsibility, understood as the need to be accountable to society for its actions. At the same time, however, the concept of social responsibility has not yet adapted to these new demands; on the contrary, it seems to be losing its ability to generate trust in the company. Today, it is, so to speak, "under suspicion". So much so that we hear many voices in business theory and practice advocating its abandonment, dragging the concept of business ethics with it. This is not the position of a discursive conception of business ethics, responsible for analyzing the conditions or assumptions underlying the credibility or social legitimacy of the company, the trust it deserves.

Keywords

Ethics • Trust • Business ethics • Social responsibility • Civil society

The publication is part of the project PID2022-139000OB-C22, funded by MCIU/ AEI/10.13039/ 501100011033/FEDER, EU, as well as of the activities of the research group of excellence CIPROM/2021/072 of the Comunitat Valenciana.

D. García-Marzá (✉)
Universitat Jaume I, Castelló de la Plana, Spain
e-mail: garmar@uji.es

A. Cortina and M. Correa Casanova (eds.), *Applied Ethics from Medicine to Humor*, Integrated Science 21, https://doi.org/10.1007/978-3-031-59009-2_5

1 Introduction

A business ethic for the twenty-first century must clarify its relationship with one of the concepts with which it is most closely related: corporate social responsibility. Today, the company is a clear example of the new protagonism of civil society, of a new distribution of responsibility for the public, as shown by the direct call for its current involvement in the fulfilment of the Sustainable Development Goals. In other words, the company is increasing its capacity for action and, along with it, its social responsibility, understood as the need to answer to society for its actions. At the same time, however, the concept of social responsibility has not yet adjusted to these new demands; instead, it seems to be losing its capacity to generate trust in the company. Today it is, so to speak, "under suspicion," to the extent that in business theory and practice we hear many voices advising its abandonment, and along with it the concept of business ethics.

This is not the position of a discursive concept of business ethics, responsible analyzing the conditions or presuppositions that underlie the credibility or social legitimacy of the company and the trust it deserves. The idea is rather to argue that social responsibility is a key concept for understanding the company, provided it is embedded in an ethical framework that, on the one hand, prevents a merely strategic and manipulative use, and on the other hand, enhances the company's public commitment as part of civil society. Let us not forget that business credibility and legitimacy (i.e., the trust that we have in the company as an institution) depends on its ability to answer for how it uses its power in the eyes of that same civil society—its social responsibility. Within this proposal for business ethics, it is necessary to design of a new generation of codes of ethics that can recover and maintain this trust by including in the code its compliance system in a verifiable, transparent, and participatory manner. Business ethics must demonstrate that the company "feels" this responsibility, and that it "means business" when it comes to responding to what is expected of it.

2 The Character of the Company: Ethics and Social Responsibility

Just as with people, organizations, administrations, companies, universities, etc., also have a character—a certain way of saying and doing, of behaving. In Greek, the word for "character" is "ethos." Hence ethics refers, in this sense, to the character of people and organizations. We say that a company is competitive, efficient, productive, etc., but we also say that it is reliable, responsible, integrated, or respectful. With these values we describe its character: that which distinguishes it from other companies. This character is achieved by making prudent and fair decisions, and by behaving in certain ways. Over time, these behaviors become habits, good practices, a source of motivation and strength to act; they shape the will and reputation of the company. There is no company or organization without character [3]. When we say "that has no character," we are saying that it does not have a defined personality, or that it does not have the courage or strength to undertake its functions. Not having character is, in fact, a character trait. Business ethics is aimed at converting character into a feature of reputation, credibility, and to generate trust.

The word "responsibility," has a different meaning and refers to a different reality. Although responsibility is often related to cause, even to guilt, its primary meaning is found in the action of *responding*. The concept has to do with our ability, personal or corporate, to answer to others for what we have done, to justify—or to adjust—our behaviors. Social responsibility refers to the company's response to society justifying that it has complied with what is expected of it—of the goods (now called "impacts") that it has produced and that legitimize its activity [9]. If it is not able to justify to society (its interest groups) the goods produced, it loses credit; little by little it is disfigured, and the moral support that sustained it is dissipated. The benefit of the concept of social responsibility is that it is now part of the language of business, which has managed to standardize what the company should contribute to society. Today we know, and therefore expect and demand, that corporate responsibility is not limited to economic profit, but also includes social and ecological profit. Its current use, however, is not free from suspicion.

In the current setting of global networks and interdependencies, States are losing their potential for intervention, thus diminishing the strength and effectiveness of legal resources to deal with the necessary regulation of action that business activity requires. States continue to be the main actors in this global framework, but they are no longer the only ones with the capacity and power to manage this new scenario. They must share their power with new global actors, such as large multinational corporations, but also global civic and solidarity organizations. This increase in the role of civil society is the ultimate reason the company finds itself faced with the need to answer for its actions and to justify its power today more than ever. The company's response can no longer be reduced to the economic benefit of its shareholders, and legal compliance with the rules of the game.

Business, as one of the basic institutions of civil society, cannot look the other way and remain silent about its role in this *new distribution of responsibility*. Simply put, greater prominence means greater visibility of social expectations on which its credibility depends. As the company's leading role increases, so does its *institutional vulnerability*. From its commitments to generate and maintain employment, to respect and improve the environment, to participate in the economic and social development of the territories in which it operates (to give just a few examples), whether or not it is legally enshrined society now expects this from a company. Whether or not it wants to, the company cannot fail to see itself as an "agent of justice" [10].

Almost twenty years ago, the European Commission's famous *Green Paper. Promoting a European framework for Corporate Social Responsibility* assured that social responsibility would be the key to making the European economy "the most competitive and dynamic knowledge-based economy in the world, capable of sustainable economic growth with more and better jobs and greater social cohesion." However, the current situation of the European economy is far from this moral high ground; instead, it could not be more disappointing. As things stand, the concept of social responsibility not only does not generate confidence, but its use by business is being called into question. There is even talk of replacing it with other less committed and apparently simpler concepts, such as *sustainability,* although this does not seem to be useful either, given its assimilation in the European context with one of the dimensions of responsibility: ecological responsibility. In view of the factual impossibility of avoiding its merely instrumental and manipulative use, some think that it would be best to abandon the concept. What has happened to bring us to this reality? The answer lies in the relationship between business ethics and social responsibility.

When it comes to understanding the value of social responsibility and, ultimately, the place that the company occupies, can occupy, and should occupy in society, we find critics who only see in this concept a progressive *substitution* of the obligations of the State, basically in matters relating to social justice. In this vision of social responsibility, the company actively participates in civil society, either by: contributing resources to a charitable and non-profit third sector through donations, sponsorships, social investments, foundations, etc.; or by extending its presence to areas where, until now, only the State had operated, as is happening in our country with the progressive commercialization of health, security, education, etc. In this replacement of the State by civil society, corporate social responsibility is expected to pick up part of the damage caused by these strategies of liberalization and privatization, however, always on a voluntary basis. Although we all know that voluntary commitments do not usually go very far if they have to face the bottom line.

A clear example of this first reading of the value of social responsibility can be found in the intentional confusion between *social responsibility* and *social action*, to which we are so accustomed. According to the European Commission's latest definition, social responsibility refers to the impact *of the company on society* (i.e., *how* the benefits are achieved and for *whom they are intended*), while social action refers to *what* we do with part of the benefits achieved. While social responsibility is understood to be the concrete obligations that the company has towards society, social action is more about philanthropy, sponsorship, and patronage. These are aspects that constitute a part of social responsibility, but do not cover its full meaning. This shift of the concept of social responsibility to one of its dimensions shows how easy it is to manipulate it, to use it for any ends. Our large companies give little acknowledgment to their responsibility for job creation or for maintaining decent wage conditions, but they talk a lot about their foundations for local development, the granting of scholarships for university studies, social aid, etc. Criticism has not been long in coming.

In the first place, we have the criticisms that are directed at the inconsequential and superficial nature of social responsibility in relation to the scarce, if not null, results of its application. Although the concept is now part of the language of business, it has hardly made any difference to the reality at the level of business. It has failed to reach small and medium-sized companies, and large companies have seen it only as a tool to improve their image. Even in this case, the profitability provided by social responsibility is far from evident [21]. But its futility increases when we understand that not only has it failed to increase good business practices, but, much worse, it has not been able to curb bad ones either. A good example is the current economic crisis and the bad practices of the large companies that caused it, yet still boast of their social responsibility policies and excellent sustainability reports.

Second, this compatibility between social responsibility and corporate malpractice is a good illustration of the merely instrumental use of social responsibility. While social action and volunteering issues occupy a large part of the corporate social responsibility agenda, the other dimensions of corporate social responsibility (employment, for example) are totally ignored beyond *marketing* actions [26]. Every day we see financial companies, which are largely responsible for the current crisis, advertise the granting of loans as a social action, when this is actually part of their function and raison d'être. The same can be said with respect to transnational corporations and their role in development—in the fulfillment of human rights, for example—where they can barely camouflage their role as authentic agents of injustice. The conclusions of the latest reports published by the United Nations Research Institute for Social Development are blunt: "As currently conceived, corporate social responsibility does not respond to development objectives."

Third and finally, the results achieved so far in the application of social responsibility seem to confirm the suspicion that it is used as an alibi to stop the demands for greater legal regulation, and for a real coercion that forces companies, especially transnationals, to change their social impacts. In this way, talking about business ethics, voluntariness, and freedom is the perfect alibi to renounce the "demand for effective public control." It is always better to recognize that we have made a mistake and apologize, than to pay fines of millions of dollars or face the corresponding criminal sanctions. Social responsibility would thus be the new sheepskin to cover the immorality of corporate actions [17].

Curiously, these criticisms coincide with those raised from neoliberal positions when they also refer to the *fraud of social responsibility*. If we are clear, as Friedman tells us in his classic article, that the company should only be responsible for "making as much money as possible in accordance with the basic rules of society, both legal and ethical customs," then the value of social responsibility is reduced to a mere instrument [7]. For this author, these concepts do nothing more than conceal actions carried out *for other purposes*. Specifically, we are faced with a strategy to achieve greater economic benefits, either through tax exemption, or through community loyalty, etc. What is disguised as social responsibility is nothing more than a decision taken in the self-interest of the shareholders or owners. Friedman does not hesitate call out these practices as hypocritical and close to deception, since, in short, they achieve nothing more than "generating goodness as a product of justified expenses for their own benefit" [1].

Behind this coincidence lies a shared vision of what the company is and can be, a vision reduced to a simple business where only the maximization of the owners' or shareholders' profits counts. An understanding of the meaning of the company that is still the usual one is observed in business schools, to the point of talking about *immoral profit strategies* [20]. This association between social responsibility and competitiveness seems to indicate that self-regulation is only understood as an instrument for the maximization of profits. This seems to be confirmed by a corporate performance that has only been able to appreciate this instrumental, purely strategic use of its social responsibility. In the above-mentioned *Green Paper,* the word "ethics" never appears, but the business argument is repeated on several occasions. Faced with this reality, it seems that the most prudent thing to do would be to abandon the concept of corporate social responsibility, and focus on the possibilities of *safe legal control.*

In my opinion, if it is not very sensible to expect companies to advance in self-regulation, it is even less so to expect the law take charge of the whole scope of the responsibilities that are proper to them. This is not only because the law loses strength (i.e., the capacity to sanction) in supra-state contexts, nor because it does not seem to be very effective in the state framework (as its ineptitude to supervise the financial markets, for example, very well shows), but also because we are dealing with two different and complementary regulations of action, as we will see later on. This is the basic task for business ethics: to discern if it is possible to understand social responsibility from an ethical perspective, thus highlighting its moral, and not only its strategic, value. Is it possible to speak of responsibility as part of the character of the company, of its way of being and doing? Only if social responsibility is at the heart of the company and is part of its character can we turn it into an asset for the company—into a moral resource.

3 Moral Resources: The Moral Contract of the Company

An affirmative answer comes to us if we stop understanding the company as a private business, and see it for what it is: a socio-economic institution. As part of civil society, the company has power to contribute to the satisfaction of all the legitimate interests involved; its own capacities for action can be used to solve problems. In this sense, promoting social responsibility means empowering individuals and their stakeholders to assume their share of responsibility, to engage in the use of this "common moral heritage" [22]. To make this common morality explicit, we must clearly differentiate between *an instrumental and an ethical use of social responsibility.* This is one of today's most urgent tasks for business ethics.

In order to understand the moral value of corporate social responsibility, and to propose an ethical use of it, we must start from the observation that there is no "zero-sum game" between moral resources and legal resources: one represents our capacity to behave morally and to use our moral knowledge in the establishment of interactions; the other, the embodiment of this knowledge in the form of legal regulations. The fact that we are calling for a greater presence of law does not mean that we are renouncing the moral demands that underlie it. The legal obligations Law of Equality has not meant that we ceased to demand and monitor an effective equality of women and men in the company; moreover, it does not displace the institutional commitment that now requires compliance. In fact, we know of many companies that comply legally, and yet are far from generating any confidence in their activities.

This second reading of the value of responsibility acquires greater significance when we clarify the existing confusion between *public responsibility* and *state responsibility*. As we see in some of the criticisms, this confusion loses sight of the assumption of responsibilities by the company as an institution, and the respective interest groups that make it up. It goes without saying that responsibility is proportional to power. Not having equal responsibility, however, does not mean having no responsibility. Talking about a new distribution of responsibility implies appealing to the strength and power of consumer organizations; to the possibilities of intervention by local government through, for example, social clauses; to the participation of workers and other parties involved in corporate governance, etc. In short, it implies recognizing the possibilities that open up when thinking about both corporate citizenship as an institution, and active corporate citizenship on the part of their respective interest groups [4].

An ethical use of social responsibility requires framing the business response within this conception of the company's commitment to the public sphere. This non-instrumental use is well-understood if we refer to the reciprocal interplay of expectations that constitutes the moral sphere. The trust placed in a company depends on whether it has been able to respond to the fulfilment, always approximate, of the expectations it generated in its stakeholders of what is expected of it. The degree of fulfilment of these expectations cannot be achieved without the participation and agreement of all the actors involved and affected. Within this reciprocal set of expectations, social responsibility is not merely a strategy for the company to improve its image, but the way in which it responds to the space of freedom it possesses. In other words, it is part of its business ethics.

From this ethical perspective, corporate responsibility can be viewed as a kind of *moral contract* [8]. On the one hand, we have the company as an organization in which certain interests converge, not necessarily antagonistic but often conflicting. As a cooperative social activity, it requires a series of material, technical, and human resources for its functioning. On the other hand, we have society, which expects a series of goods from the company, which will ultimately justify the use of resources and the asymmetrical distribution of burdens and benefits that characterize any type of business structure. These goods that are not reduced to economic profit, but also include, for example, respect for the equal dignity of all the people involved, attention to and improvement of the quality of the environment or participation in territorial development. We speak then of shared value.

In other words, the business activity induces a series of expectations in the internal and external groups involved and affected by it. These expectations refer to its corporate mission, to the activity it carries out, and to how it carries it out. If society—the public opinion formed through the different interest groups—perceives that the company responds and fulfils these expectations with sufficient approximation, it then provides the necessary confidence, and grants, for example, sufficient credit to achieve and guarantee a good working environment, or loyalty when buying a product. When we speak of a moral contract, we are not referring to a legal contract, but to this reciprocal interplay of expectations that underlies it.

While the strategic use of social responsibility refers to a company model linked to one interest group (i.e., the owners or shareholders), the ethical use requires a plural model of a company shaped as a socio-economic institution whose legitimacy depends on the possibility of a free and voluntary agreement between *all the* interest groups involved in its activity. From this understanding, corporate social responsibility does not depend on a calculation of results carried out by the managers; however, the validity of the contribution made depends on the acceptance of these results by *all* those involved in a free and open dialogue, carried out under equal conditions of participation. These are two different business models, and are antagonistic most of the time.

The idea of dialogue and subsequent agreement between all the stakeholders involved in and affected by the company's activity thus becomes an *ethical framework*—a horizon of meaning from which to construct reasons to assess the legitimacy of an institution, and so justify or criticize its actions. The distance to the moral horizon of the possible agreement between all the participants in conditions of equality finds its empirical basis in the trust or distrust the company deserves. This is the core of a business ethics that has its theoretical basis and normative foundation in a renewed ethics of discourse. In the form of a principle, it can be synthesized with the following formulation:

> a business decision or action, as well as the company as an institution, can be considered just or correct if it can count on the free and voluntary acceptance, under equal conditions of participation, of all the interest groups involved and affected by the company. [8, 24]

In short, the ethical use of corporate social responsibility requires framing the responsibility within this moral agreement that underlies the legitimacy of the company, as a response to the achievement reached in the satisfaction of the expectations at stake. The fact that social responsibility is not justified by the calculation of consequences but by the agreement reached does not mean that this ethical use is of no consequence. On the contrary, as a moral resource it provides us with a better relationship with stakeholders, an increase in trust and reputation, lower coordination and transaction costs, a good working environment, etc. Of course, we must always bear in mind that the profitability of ethics for the company derives from its character as a moral resource, not the other way around.

4 Ethics in the Institutional Design of the Company

The image of the moral contract, as an expression of the reciprocal recognition that all interaction presupposes, allows us to visualize the ethical perspective in which an ethical use of social responsibility must be inscribed. From this perspective, the objective of a company that deserves the label of responsible cannot be other than the agreement of all the actors involved in its activity. There is neither a business interest identifiable with one of the groups—no matter how powerful it may be—nor is there a corporate interest beyond the satisfaction or fulfilment of all the legitimate interests at stake. The good contributed by the company can be located in the benefit, but this statement is completely empty if we do not immediately clarify *what* the benefit consists of, for *whom* it is intended, and *how* it has been achieved. The agreement will only be possible if this production of goods reaches all those involved and affected, a question that we can only know through their *participation.*

In this way, a clear distinction is made between *business ethics* and *social responsibility.* The ethical perspective is positioned at a normative level and establishes the conditions from which it is possible to speak of a fair dialogue or a moral agreement between the different parties involved. Social responsibility, on the other hand, refers to the set of actions, decisions, and policies that make up the response offered by each specific organization to the demands and requirements of its corresponding stakeholders. That is, it determines the degree of approximation and factual commitment to the ethical horizon of dialogue and possible agreement, aspects that constitute the ethical foundations of trust. We must not confuse the criteria for action with the conduct to which they may give rise.

However, business ethics—as applied ethics—cannot stop at defining and justifying the moral resources that business, as part of civil society, possesses; nor should it stop at determining the meaning of an ethical use of social responsibility. It must go a step further and also justify that what is desirable is also possible, that we are not facing an "unreality." And this is where the problems begin. The price paid for introducing the participation and agreement of all interest groups as a criterion of moral validity is none other than its excessive abstraction, barely attenuated by this empirical reference to the trust or distrust generated. It seems, therefore, that fleeing from a fraudulent use of corporate social responsibility, we have ended up with a concept that is far removed from reality, superfluous, and useless for explaining the role of the company in civil society. An idea, by the way, easily manipulated.

These are problems that become much more complicated in the face of the complexity and anonymity of transnational corporations, where it is always difficult to find out *where, who* and *how* they have the power to make decisions. The dimensions of these global companies do not even allow us to think about the participation of all the groups involved, not to mention the future generations directly related to sustainability. To repeat the approaches of industrial democracy under these conditions is not only naïve, but can be counterproductive as the normative and critical perspective can be lost. The question is obvious: how can we talk about agreement without talking about the participation of all those involved and affected?

A business ethics based on a renewed ethics of discourse can overcome these problems if we pay attention to a basic question already mentioned: the business cannot be conceived or explained as a "private affair," as a legal contract between individuals. It is thus forgotten that we are dealing with an institution and that it therefore requires a structure, values, and norms that determine the necessary coordination of action among the different interest groups. And, most importantly, it ignores the fact that this structure can be modified, from the type of governance to ownership, including the criteria for the distribution of burdens and benefits. If business ethics is concerned with the ethical foundations of trust, it must make progress in defining an *institutional design* that allows and enhances the ethical use of social responsibility, i.e., enables the use of moral resources without endangering the very existence of the company. In my opinion, we can advance in this design if we take two basic steps [12].

First, this new institutional design requires another type of theories of a more empirical nature, capable of specifying these normative criteria of participation and agreement. This is the case of the *stakeholder* theory or approach, whose methodology allows us to *operationalize* the idea of agreement by helping us to define each of the groups involved with the company, as well as the interests that involve them, and the conditions in which they are fulfilled [14]. This theory is always directed towards a basic question: organizations are composed of a plurality of interest groups on which their credibility or social legitimacy depends. This objective always implies recognizing a double direction in these relationships, as we see in Freeman's classic definition of stakeholder: "any group or person who can affect or be affected by the achievement of the company's objectives." In the same way that a stakeholder can be affected by the decisions, actions, and policies of the organization, it can also affect the smooth running of the company [6]. Without this help, it is not possible to delimit and manage the idea of the moral contract.

Second, the moral requirement of participation and agreement represents a horizon for action, an idea that defines the morality of the company, a point of reference, so to speak. If it is to be credible (and therefore effective), any measure adopted that gives effect to social and ecological responsibility requires that the company, in each specific situation, *make public* the measure it has adopted. The requirement for dialogue and agreement is still there, but in each specific case it becomes a source from which it is possible to *provide reasons that justify* the position adopted by the company, for example, in the determination of the groups that compose it. Institutional design theories work with this idea by referring the capacity to produce trust to this potential discursive justification [15, 16].

From this approach, what is decisive in order to speak of a responsible company, of an ethical use of social responsibility, is that it demonstrates the steps it is taking to lead its management towards this model of company that has a foundation in participation, in dialogue, and in possible agreement. This ability to justify itself is already part of the moral contract on which the company's credibility (i.e., its responsibility) is based. It is not just a question of a willingness to be honest, but also of this willingness to acquire the status of a *public commitment*. This is the only way for responsibility to become a corporate asset, and a key resource for generating trust.

This approach to *public justification* is no recent "invention." Just as a trustworthy person is someone who exposes himself to continuous and scrupulous scrutiny by others, so it is with institutions. The proposal to use transparency to show goodwill was very well systematized by Kant as a *principle of publicity*, understood as the requirement to "make public," to give publicity so that all those possibly involved and/or affected are aware of it. It is about asking oneself: if the motives that underlie my decision were in the public domain, could do what I do? Negatively, this principle says: "All actions relating to the right of other men whose principles cannot bear to be published are unjust" [18]. Silence is the fundamental anchor of distrust.

But transparency alone is not enough to ensure this ethical use and build trust. Openness is necessary, but it has only one direction and depends too much on the power of the company. In his time, thinking precisely about the application of ethics, Kant goes much further and offers us a positive definition of the principle of publicity where consent or agreement on the results achieved is sought. Today, we can deduce this need for publicity from the monitoring and control of *all* stakeholders, that is, from its inclusion in an institutional design that responds to the efforts made by the company to approach participation and free agreement [11].

For a business ethics that not only wants to define a criterion of justice for the company, but also wants to achieve effectiveness, the solution is to update this principle of publicity and to implement it in the institutional design of the company as a corporation.

This brings us back to the basic meaning of the word "responsibility," i.e., the ability of the company to be accountable to society for what it does or does not do. We do not have a "measuring rod" to decide, in each concrete case, the responsibility of a company; it always depends on the social and economic conditions, on the determined contexts, on the available resources. But we can design communication and participation mechanisms so that we have a *rational basis* on which to base our perception and assessment and, with them, our trust or distrust in business activity. In short, we can design an *ethical infrastructure* that supports the ability to define and use the moral resources that move in the company.

It is this ethical infrastructure that must account for this condition of publicity and that establishes the difference between an ethical use and an instrumental use of social responsibility. Only from these conditions is it possible to speak of responsibility as a moral resource, and avoid the criticism of instrumentalization and deception. Hypocritical action does not stand up to this demand for publicity, understood as the sum of transparency and participation. On the contrary, a company that relies on these principles has broken distrust, and provides the appropriate framework to motivate support for the company, and trust in its decisions and policies.

5 A New Generation of Codes of Ethics and Conduct: Ethics and Compliance

Organizational codes of ethics have been the traditional mechanisms used to understand the ethical commitment of the organization, and its commitment to social responsibility; more recently, social responsibility reports or sustainability reports have been added. However, as we will see below, we need to integrate them into a broader organizational design that includes, according to the ideas outlined above, not only transparency but also participation.

The construction of this ethical infrastructure, the application of the principle of publicity to institutional design, admits multiple possibilities. We can specifically recognize four fundamental steps when speaking of business ethics, of an ethical management of corporate social responsibility. The first two steps of this ethical infrastructure correspond to the principle of transparency, the last two to the principle of participation [8].

The first step in this generation of trust is the elaboration and publication of *codes of ethics* and *conduct*. These are formal documents where we find an explicit statement of the values that should guide the conduct of employees and managers, thus promoting good practices and assessing the character and personality of the organization [19]. Its function is, therefore, twofold: internally, to formalize the values and decision criteria that define the organizational culture itself, as well as the conduct expected of its employees; externally, to manage the reputation of the organization. In other words, it presents us not only with the values that define the character or ethics of the organization, but also the commitments that the entity is willing to assume to create this common will and the conduct necessary for its realization.

The importance of codes of ethics grows with the global nature of many of the problems that organizations must face today, whether they are companies, administrations, or civic and solidarity organizations. Faced with these global and multicultural challenges, law (always limited to state borders) and tradition itself (which depends on each particular culture and religion) can do little. Ethical codes do, however, present values, commitments, and conduct that are specific to the organization wherever it operates, thus showing a character, a way of being and doing. In short, an ethical commitment must be in line with the moral values expressed in international guidelines and agreements, such as the Global Compact or the Sustainable Development Goals.

Second, there are the *social responsibility reports* or *sustainability reports* which, by using a standard methodology—as is the case with the *Global Reporting Initiative*—allow the company to publicly present the results achieved or, as it is now called, the impacts generated. These reports must contain both the commitments made and the management approaches taken to fulfil or satisfy the legitimate interests involved and affected by its activity. The last part deals with measurable and verifiable indicators, so that monitoring and control can be carried out.

Although these reports inform us not only of management policies and strategies but also of the results in the form of indicators, their preparation by the communication or corporate social responsibility departments—or, worse still, by specialized consultancy firms that are also responsible for verifying the information provided—does not achieve the desired credibility. Both the monitoring and control of compliance with the commitments of the code of ethics and the veracity of the information provided in the responsibility reports must be open to stakeholder participation. In this way, responsibility becomes co-responsibility, a common task, a will shared by all stakeholders.

The proposal advocated here is not to abandon the first two steps in the generation of trust, but to integrate the need for transparency with the need for participation, so that these two instruments of accountability are supported by the participation of all stakeholders involved in and affected by organizational activity. To achieve this objective, we must include in the code of ethics the compliance mechanisms as part of the code itself, and not as separate parts. In this sense, we are talking about a renewal, indeed, a new generation of codes of ethics, precisely because they incorporate participation not only in the definition of commitments, but also in the monitoring and control of their fulfilment.

Third is the basic function of the Ethics and Corporate Responsibility Committee, as a response to the demand for deliberation and participation of the different stakeholders, that is, as a "public space" within the corporations themselves. This committee functions as a place for participation where all stakeholders are represented and where monitoring and control of social responsibility policies can be carried out, from the preparation of social responsibility reports to the coordination of communication channels for the notification of alerts and irregularities. Its function is threefold:

- *Advise* on issues related to the interpretation and application of the code of ethics.
- *Resolve* suggestion notifications, alerts, and complaints made through the ethics hotline.
- *Promote* information, and employee and manager training in the ethics and compliance program.

Its composition must respond to the organization's main stakeholders, always aware that it is to assist in the management and compliance with the values and standards expressed in the code of ethics and conduct. The representation of the different groups in the Ethics Committee can be done in various ways (e.g., based on the reliability or reputation of the elected representatives), as long as the criteria for representation are themselves public. Trust in the committee will depend on the trust that its members are capable of generating [8]. Its functioning is framed by the principles of independence, transparency and consensus. These principles mean that the committee must always have representation from external stakeholders and that, of course, they have no relationship whatsoever with the boards of directors.

There is one more step. The participation required in this proposal for company ethics should not be limited to a small committee which, although it provides the presence and voice of internal and external stakeholders, does not constitute the voice of *all.* We must establish channels of communication that allow the participation of all those who wish to do so, always focused on the fulfilment of the ethical commitments adopted. Communication should not be limited to the reporting of bad practices. Above all, an ethical culture demands the involvement of employees to both: alert, prevent, and detect situations and conflicts of interest that may lead to non-compliance and damage the reputation of the company; and propose and suggest improvements in and the development of good business practices. The same applies to external groups, users, clients, suppliers, etc. The aim is to progress in an ethical culture where the implementation of ethics is a shared responsibility.

This is the role of the *ethics hotline*, understood as a communication channel for the expression of suggestions, alerts, and complaints. In any format (post, e-mail, website, etc.), it is always a matter of communicating both breaches of the code of ethics and suggestions for improvement. Of course, communication must always be confidential, but never anonymous, as anonymity is the opposite of the requirement for publicity. However, the necessary protocols must be established to ensure confidentiality and avoid any possible reprisals. Suggestions, alerts, and complaints should therefore be directed to the ethics committee, and not just to the ethics and compliance officer.

In short, we are faced with different progressive steps to generate trust in corporate social responsibility. We must provide reasons that achieve the conviction that this is an ethical management of social responsibility—not merely an instrumental or strategic use. These steps are not independent, rather they constitute a system in the sense that they refer to each other when we think about managing trust. This is why they should be included in the ethical code itself: they are commitments made by the company to society and, therefore, a guide for decision-making and good practices. The structure of this new generation of *codes of ethics* and *conduct* should include, as a minimum, the following sections:

(1) Letter from the Chairman on the importance of the code and management's commitment to it, highlighting the integrity and exemplarity required for its operation.
(2) The basic moral values that define the company's character and that should be part of its vision and mission.
(3) The principles of action derived from these values and the requirement of the respective conduct, specified in thematic areas so that they are verifiable.

(4) The ethics and compliance system, i.e., how the company designs both the preparation of responsibility reports and the composition and functions of the ethics and corporate responsibility committee, and the functioning of the participation channel or ethics line [13].

These mechanisms for an ethical management of social responsibility must be understood as progressive steps that the company must take if it wants to turn it into a moral resource, i.e., a business asset. We have insisted that a responsible company is one that moves towards inclusion of all the legitimate interests at stake, configuring the necessary structures and policies so that transparency and participation move the dialogue and the possible agreement between all involved. Within this framework, designing ethical codes as mere declarations of intent is a very easy thing to do, but without the reinforcement of the above steps, they only serve to increase mistrust.

A good example of the current importance of these instruments for ethical management of social responsibility can be found, in the Spanish context, in the amendments to the Criminal Code in 2010 and 2015. These amendments already speak of the "process of formation of the will of the legal person," of the company as a potential subject of crime by stating that "legal persons shall be criminally liable for crimes committed in the name or on behalf of the same," and recognizes the value of this type of mechanisms by considering as mitigating circumstances, or even exonerating circumstances, the prior existence of "effective measures to prevent and discover crimes." Following these guidelines, and especially as part of the fight against corruption, large companies have made progress in implementing these compliance programs. In this way, ethics helps the law prevent and control crimes; and the law, in turn, gives greater significance and importance to the mechanisms, which, based on voluntariness and commitment, are proper to civil society.

Thus we speak of complementarity, and not in any way of substitution. While in law it is the sanction that is the main motivation behind this legal call for self-regulation, in ethics it is the inner conviction that we must act in this way, our obligation derived from the legitimate expectations placed in the company.

Ethics and law are complementary paths that have their own logic and their own laws of operation. While law tries to positivize these values and norms in order to convert them into *legal resources*, thus achieving external coercion and a greater guarantee of security (this is at least the intention), business ethics tries to make *moral resources* explicit and manage them, while maintaining the voluntariness derived from our freedom of choice and action. But voluntariness does not mean discretion or arbitrariness, which is a mistake made by many critics of business ethics. The others involved in the business activity "expect" us to behave in a certain way, they trust us to respond in a certain way, and we know it. These reciprocal expectations of behavior form moral knowledge capable of becoming resources for achieving interactions and resolving conflicts. If we act according to what is expected of us, we generate trust and with it the *moral credit* necessary to continue establishing relationships with our stakeholders. Otherwise we increase *moral hazard*, the possibility that the company does not act as expected and thus loses this moral support in the form of a bad reputation, loss of customers, or increased labor costs, for example [25].

In short, this new generation of codes of ethics goes beyond the current compliance systems, which are in the hands of large consulting firms and always limited to the legal sphere, concerned only with *criminal risks* [23]. The law must be present for crime prevention, but compliance cannot be understood without ethics and social responsibility, without transparency and participation. For business ethics, it is this ethical framework that allows us to manage both resources and moral risks.

6 Conclusion: The Friendly Company

A company that acts within this ethical framework does so out of conviction, not out of coercion or convenience. Only when it acts because it believes it should do so does its social responsibility become an asset, what we call a *moral resource*; only in this way is it credible and, therefore, capable of generating trust, of reducing the moral risk that all human relationships, economic or otherwise, entail. When we speak of an ethical company, we are talking about the normative horizon of dialogue, and possible agreement of all the stakeholders involved in business activity. Social responsibility is the road travelled; it is the response given to this moral requirement, and the results of the commitment made. In the distance that exists between the two, the trust that the company is capable of generating is based.

But business ethics is not limited to the construction of an ethical infrastructure that makes participation and agreement possible; it must also be concerned with activating it, with setting it in motion. It would be an artifice to think that by having a good institutional design we have already guaranteed ethical behavior, or that by having well-defined values and norms we have already turned this moral knowledge into a reality. What a good ethical infrastructure must do is not merely prevent bad practices. It must, in particular, allow and promote good practices, make them become the norm in the company. To this end, we must consider a key element in the composition of this moral knowledge: emotions and their perception as feelings. It is the motivation of the action towards which we are heading, the fact that the company "feels" that it must act in this way, that it is "serious."

The task of business ethics does not end with the creation of an ethical infrastructure that favors the use of moral resources; it must also advance in the creation of a corporate culture that, to paraphrase Ortega, constitutes the set of living ideas, beliefs, desires, and attitudes that give meaning to its activity. A renewed ethics of discourse thus becomes an ethics of cordial reason, a gamble for procedures to incorporate feelings, for affection to be the other face of justice [5]. Now, does a cordial civic ethics have its institutional translation, its application to the company in our case? Like the rest of applied ethics, it is easy to give an affirmative answer if we take into account the etymological root of the word "cordial": with the heart.

First of all, regarding the emotions and affections to which the use of the word "heart" refers. Often, when we refer to companies, we say that "they have no heart," in the sense that they have no feelings, they are insensitive, rather cruel, they have no humanity. The question is simple: if we had a choice, would we work for a company with this character? Would we buy their products? Would we let them be our suppliers?

In business, as in any field of human relations, emotions play an essential role of motivating action, and reinforcing or weakening the strength of decisions, even influencing and predisposing towards a particular decision or behavior. In the case of moral emotions, these are reactions to what we expect from others, reactions to the feelings and attitudes they have towards us, how they treat us, whether or not they recognize our work and our worth. These emotions involve moral knowledge, they have a cognitive content that can be taught and learned. Knowing how it works, what provokes satisfaction or pride of belonging or, on the contrary, indignation, resentment or anger, is not trivial. We are not talking about chemical and hormonal responses, easily manipulated, but about moral feelings and the reasons behind them. If we were able to understand the moral emotions that move in the company, we would immediately notice the motivational heterogeneity that defines, also there, all human action. To insist on explaining all behavior from the point of view of selfishness, the profit motive, and competitiveness is naïve, and hinders the generation of value for all the stakeholders of the company [2].

Secondly, it is also common to hear that social responsibility should be part of the company's DNA. In my opinion, I think it would be better to think about this cordial ethic, and affirm that it should be in the *heart*. The heart is a muscle that must be exercised at all times, it is the will to do things well; whereas genetic inheritance is neither the fruit of our will nor does it have to be activated. From this point of view, ethics has to do with *courage*, bravery, and the decision to comply with the commitments set out in the code of ethics; with the *sanity*, prudence and good sense necessary to carry them out; with the *agreement* of all its stakeholders, internal and external, as a horizon for action; and, finally, with *remembering* that the company depends directly on the resources provided to it by society, and if it is not capable of being accountable for their use, it will disappear. From this cordial ethic, business ethics is not a mere ornament or cosmetic, it is rather a necessity, a condition of possibility of the meaning and existence of the company. To think of business ethics as an asset and as a factor of social innovation, or to think of it as a new social tax depends on these coordinates.

References

1. Bowie N (1991) New directions in corporate social responsibility. In: Business horizons, pp 56–65
2. Conill J (2004) Horizontes de economía ética. Tecnos, Madrid
3. Cortina A, Conill J, Domingo A, García-Marzá D (1994) Ética de la empresa: claves para una nueva cultura empresarial. Trotta, Madrid
4. Cortina A (2005) Citizens of the world. Alianza, Madrid

5. Cortina A (2007) Ética de la razón cordial. Nobel, Oviedo
6. Freeman RE (1984) Strategic management. A stakeholder approach. Pitman, London
7. Friedman M (1990) The social responsibility of business is to increase its profits. In: Hoffman WM, Moore JM (eds) Business ethics. McGraw-Hill, New York, pp. 153–157
8. García-Marzá D (2004) Ética empresarial: del diálogo a la confianza. Trotta, Madrid
9. García-Marzá D (2007) "Responsabilidad social de la empresa: una aproximación desde la ética empresarial. Veritas J Philos Theol 17:183–204
10. García-Marzá D (2009) Agents of justice? La responsabilidad social de las empresas, In: Cortina A, Pereira G (eds) Pobreza y libertad. Tecnos, Madrid, pp 193–211
11. García-Marzá D (2012) Kant's principle of publicity. Kant-Studien 103:96–113
12. García-Marzá D (2013) The ethical dimension of institutional design. In: Gonzàlez E (ed) Ética y gobernanza: un cosmopolitismo para el siglo XXI. Trotta, Madrid
13. García-Marzá D (2017) From ethical codes to ethical auditing: an ethical infrastructure for corporate social responsibility communication. El profesional de la información (EPI) 26(2):268–276
14. González EE (2007) Stakeholder theory. A bridge for the practical development of business ethics and corporate social responsibility. Veritas II(17):205–224
15. Goodin RE (Comp.) (2003) Institutional design theory. Gedisa, Barcelona
16. Guy Peters B (2003) El nuevo institucionalismo. Gedisa, Barcelona
17. Hernández J (2009) El negocio de la responsabilidad. Icaria, Madrid
18. Kant I (1987) Perpetual peace. Tecnos, Madrid
19. Lozano JF (2004) Códigos éticos para el mundo empresarial. Trotta, Madrid
20. McDonald D (2017) The golden passport: Harvard Business School, the limits of capitalism and the moral failure of the MBA Elite. Harper Collings
21. Moreno JA (2014) To overcome rhetoric. Dossiers EsF 14:4–11
22. Offe C (2004) Las nuevas democracias. Hacer, Barcelona
23. Peterson EA (2013) Compliance and ethics programs: competitive advantage through the law. J Manage Govern 17(4):1027–1045
24. Ulrich P (1987) Transformation der ökonomischen Vernunft. Haupt, Stuttgart
25. Villafañe J (2004) The good reputation. Claves del valor intangible de las empresas. Pirámide, Madrid
26. Vives A (2001) A critical look at CSR in Latin America. Cumpetere

Ethics of the Professions: Virtues, Vocation and Civil Society

6

Agustín Domingo Moratalla

Abstract

In the age of digitalization, the transformation of professions is accelerating. The changes that have occurred in human activities have a direct impact on the way professional life is understood, trained, managed and led. The aim of this chapter is to analyze these changes from an ethical perspective in order to update the basic concepts of any ethics of the professions. The goods, practices, institutions, virtues and traditional concepts of work ethics need to be updated to respond to the new challenges of management applied to companies, public administrations and institutions in general.

Keywords

Internal goods • External goods • Institutions • Virtues • Professional career • Training • Theory and practice

1 Introduction: Ethics, Deontology and Applied Ethics

When we study the "Ethics of the Professions" within contemporary Moral and Political Philosophy, it is important to clarify why we focus on an Ethics of the Professions (EP) and not on a "Professional Deontology" (PD). These are two terms that are sometimes presented as interchangeable. In the field of everyday morality are very easily confused.

A. Domingo Moratalla (✉)
Universidad de Valencia, Valencia, Spain
e-mail: agustin.domingo@uv.es

A. Cortina and M. Correa Casanova (eds.), *Applied Ethics from Medicine to Humor*, Integrated Science 21, https://doi.org/10.1007/978-3-031-59009-2_6

To avoid these confusions, we must remember that the name "deontology" comes from the Greek word "deon" which means "duty" and, therefore, deontology includes the set of rules, obligations and duties required of those who exercise the same profession. The term "ethics" also comes from Greek and refers to "ethos," which can be translated as "character," or "second nature". When– applied to a certain group—to the set of uses or "practices" carried out by those who exercise the same profession. In the ethics of a profession, we ask ourselves about the relation that these uses or practices have with personal conscience, with the general good, with values, or with ideals of life. These ethics of a profesion represents, for instance, the good thing to do given an specific situation faced by the professional. It also may represent what is interpreted as morally good on the professional field of expertise.

This takes the form of three philosophical tasks. First, to discover the values that appear or are at stake in these uses or practices; second, to clarify these values in order to analyze the possibilities they offer for the pursuit of excellence in a given profession; and third, to give an account and reason for the decisions we make on a daily basis when we face the problems involved in the exercise of our profession.

This contrast between "values, goods, practices, and virtues," on the one hand, and "duties, norms, or obligations," on the other, cannot be seen as confronting or opposing each other. When we analyze the different professions of a society, we have to observe in them two complementary spheres. First: Norms, rules, and duties will always be necessary in a profession, secondly, the specific ethics of a profession are always at the service of the goods that give that profession meaning. We need to take into consideration that norms, duties, and obligations are necessary, but not sufficient.

The PD organizes minimum duties that can be demanded of all professionals in a group; without it there would be neither universality nor equality of requirements; also describes and presents what is binding and obligatory for all. The EP it questions the meaning of the profession in each of the professionals and, therefore, demands that the rules, duties, and obligations "are motivated," that is, that they are situated within an intentionality and a purpose with which they gain visibility, plausibility, legitimacy, and meaning.

A distinction made by Ricoeur ([16]: 47) when analyzing medical ethics and judicial ethics is useful: He used the analogy of a riverbed to describe the set of rules of a profession, and wanted us to think of ethics as the two directions of "upstream" and "downstream":

(a) Upstream, we should ask ourselves about the origin, the meaning, and the source from which the norms flow, as if we were asking ourselves about the sources from which the river is born. While the PD designates the current of the river (i.e., the norms of a profession), the EP poses the question of the origin, the place and the meaning from which the norms are born.
(b) Downstream, the rules do not apply themselves. They need professionals who discover new opportunities, decide in contexts, circumstances and situations that demand the realization of the good that the rules serve.

From this perspective EP helps us in two ways. On the one hand, it helps us to base the norms when we go "upstream," that is, to give account and reason in increasingly plural and complex societies. On the other hand, it helps us to "apply" our duties and to train ourselves in the *phronesis* or practical wisdom that we need when we put the norms into practice. With this double function which EP performs, we integrate what Karl Otto Apel's ethics of discourse calls part A of Ethics, or the "moment of foundation," and part B of Ethics, or the "moment of application."

Sometimes, instead of using the term EP we use the term *Applied Ethics*, to reinforce the need for prudential and deliberative reason in the praxis, application or historical modulation of professional life. The term is used in the plural, *Applied Ethics*, to designate ethics in a given professional field, so that EP and Applied Ethics can be synonymous terms. In the present work, EP has a general sense and its reflections are useful for the different professional or vital fields of ethics where we refer to Applied Ethics [5, 7].

2 Work and Virtue in the Digital Age

2.1 Thinking About Work: Jobs, Professions and Careers

For a long time, "the professions" were studied exclusively from the perspective of history and sociology because they were concerned with the evolution of the relationship between human beings and nature, and the way in which a society is structured and organized. EP has burst into the social sciences as a whole because, with the acceleration of social change, not only have new occupations, jobs, and trades emerged, but also new ways of organizing, regulating, and socially controlling these activities. For this reason, when we approach EP, we must distinguish between three uses.

(a) An old and strict use of the term "profession" applied exclusively to people in medicine, law, or theology.
(b) A broader and more modern usage according to which people with certain scientific knowledge and training (e.g., economists, engineers, lawyers, architects, or teachers) can be called professionals.
(c) An imprecise and colloquial sense that recovers the meaning of "professional" for that worker who seeks excellence and permanent improvement in his trade.

When we analyze professions today in the whole of human action, we discover that we need a *work ethic* that helps us to outline what we understand by profession. Here we should reevaluate the reflections on work and the arts, on trades and occupations, and distinguish servile occupations that required greater physical strength and contact with nature, from non-servile occupations that require less manual labor and more intellectual work. We can also rethink the whole of human activities from two perspectives: the perspective of necessity and the perspective of freedom.

For example, if we think of human activities from the perspective of *necessity*, then occupations, trades and professions are not chosen: we come to them simply as a means of life, as a resource with which to "make a living." However, if we think of human activities from the perspective of *freedom*, then occupations, trades and professions are the fruit of our conscious, responsible and personal choice. In this case they are not a "means" of life, but that type of activity through which our life has a meaning.

This consideration is more important than it seems at first sight, because professions are not simply jobs or occupations that we do, but something that *determines our identity*. Healing, teaching, or legal advice are not mere occupations or jobs, but are activities that shape the identity of the person who performs them. We don't usually say someone "has medicine/law as a profession." We say someone "is a doctor," someone "is a lawyer," someone "is a teacher," and so on.

This leading role in the configuration of identity means that professions are not a minor or small problem in the moral configuration of society. We are concerned about EP because the identity of the citizen is conditioned by the profession he or she has chosen or by the occupation in which he or she is employed.

In the work ethic there is an important tension between the *world of employees* and the *world of professionals*. The key is the internal relationship they have with time and the extent to which this shapes their identity. For example, the expression "having a job" describes an external relationship of the worker with time, in the sense that he/she occupies certain hours of his/her life carrying out an activity or job. He/she uses "part" of his/her time for others in the institution or company. Their work is perfectly evaluable and measurable from an "external" perspective in terms of time. In fact, there are workplaces that measure their employees' time with clocks.

Unlike those who aspire to *have a job*, there are people who aspire to *exercise a profession*. In these cases, the person's relationship with time is not purely external or instrumental, and they feel restless, called, or vocationally committed to a certain activity. The activity of a doctor, a lawyer, a judge, or a teacher cannot be measured with a clock and the number of hours spent can be calculated. Surgery, law or architecture, for example, are not simply activities in which people spend their time in order to earn a living. They are activities in which people "spend their time."

This difference between "spending time" (external relationship with time) and "spending time" (internal relationship with time) is important in order to enter into a EP that is not necessarily an employee ethic and that cannot be dissociated from a work ethic. In the moral configuration of citizenship, it is different to assume work and the challenges of digitalization in a way that is external to our personal identity (having a job) than in a way that is internal to an identity in which work plays an important role (exercising a profession).

When we consider work ethics as a *career*, there are two ways of understanding it. There is a strictly utilitarian and individualistic way of understanding the profession, in which the professional is situated and projected in time in terms of competitiveness, struggle or success, and who reads his activity in terms of triumphs or failures. In contrast to this utilitarian approach, there is an Aristotelian approach where there is also a projection in time, but in terms of talent, training and self-giving. While the first approach supposes a fragmentary conception of the profession because the worker lives obsessed with success, and focuses all his activities from work as "nec-otium" (without leisure), the second one supposes an integral form of the profession in which there is also time for the family and "otium." The first implies a fragmentation of life because the profession is seen as an activity for promotion, prestige, and advancement ([3]: 106 ff.).

In the EP, we analyze the consequences that these two ways of understanding the profession as a career may have, and we propose the need for a reflexive or critical concept of career. In contrast to the strictly individualistic approach that privileges the quantitative and vertical dimension of the profession, we propose the need for an empowering approach that considers the project in terms of co-responsibility. In a professional career, there are not only demands on oneself, but also on others; there is not only a horizon of individual responsibility, but also of community responsibility. For this reason, from the perspective of EP we can read professional careers not only in terms of "being the best" but also in terms of "continuing to do better." The obsession with "being the best" would lead us to professionalism understood as "careerism," and would return us to a corporatist society, that is, one of corporations or groups of egoists where professionals think only of themselves and forget about others. This is one of the main risks that we must avoid and that is why a EP must distance us from contributing to this reductionist and simplifying reading of the profession.

2.2 Vocation and Profession in the Performance of Duty

In his book on *General Ethics of the Professions*, Professor Hortal recalls that for a long time there has been a tendency to understand every profession as a kind of priesthood; the doctor began to be understood as a "priest of health," the judge as a "priest of justice" and the military as a "priest of arms" ([11]: 38, [8]). With this linkage it is easy to wonder about the "vocation" that one has to carry out a certain activity or occupation [18]. In this way, professions can be studied with the religious metaphor in order to analyze to what extent there is a vocational, natural, or structural element in the work that is carried out.

Read from the Protestant ethic, this religious metaphor presents vocation simultaneously as a divine calling and human profession (*Beruf* is the term introduced by Luther). It is easy to analyze vocation in terms of calling, dedication, devotion or consecration of one's own life to the occupation in which one is engaged. While from the Protestant perspective profession is read as a response to a divine call (*vocatio*), there are other perspectives where vocation is seen as a basic concern that motivates one to choose a job and to spend one's life in it.

In translating the semantic potential of the religious metaphor, we must be careful to recognize that not every *professional ethos* is rooted in Luther's and Calvin's reading of the professions. Moreover, this translation of the semantic potential of the religious metaphor into the work ethic may force a process of predestination or sacralization of the professions. It would not be the first time that this has happened, especially when the oldest professions were considered as a kind of "lay priesthood."

Although we use this religious reference, we must also be aware of the changes brought about by its application. In his study entitled *The Protestant Ethic and the Spirit of Capitalism*, Weber [19] describes how, after Luther, one no longer has to change one's occupation in order to be saved, because every lawful occupation or profession has the same value before God. With the Protestant ethic there is a levelling of all forms of work and the fulfilment of all the duties that are appropriate to each occupation. With the spirit of capitalism a certain *professional ethos* develops which links work with accomplishment, enrichment, and the culture of success.

This *professional ethos* of Protestantism continues to be maintained with the new forms of capitalism, and it is forgotten that other readings of the work ethic are possible. Professor Hortal ([11]: 42) indicates that this *professional ethos* can be reconstructed from other different philosophical or religious roots. In fact, the reconstruction that he proposes incorporates elements of the Aristotelian communitarianism, which he employs to integrate analyses of other secular and religious traditions. At the end of the chapter he devotes to delimiting the concept of profession, he points out that a profession can be defined as an *occupational activity* that has the following characteristics:

(a) In an institutionalized form, a *specific service* is provided to society.
(b) By a group of people (the professionals) who *dedicate themselves to them in a stable way*, obtaining from them their *livelihood.*
(c) Forming with other professionals (colleagues) a *collective that obtains or tries to obtain monopolistic control* over the practice of the profession.
(d) They enter the profession after a *long process of theoretical and practical training*, on which depends the accreditation or license to practice the profession.

Where is vocation in these occupational activities? Does it make sense to continue to ask ourselves about vocation in an ethics of the professions? As we have pointed out in the study of the managerial function, the question of vocation can be the key to reconstructing the ethical dimension of the occupation or the work we do. Without the question of vocation, the profession is reduced to the fulfillment of a series of duties or the performance of a certain task within an organizational chart, institution, or system. In this context, it is useful to recover the close relationship with which the hermeneutic and raciovitalist tradition propose the vocational dimension of professional life.

Let us remember the importance that Ortega and Gregorio Marañón attach to vocation, not only in professional life as a whole but also in reference to personal identity. More than procedures, sanctions, or rules, society needs professionals with a vocation who have faith in what they do; for those who have a vocation, rules and regulations will be superfluous, and for those who do not, they will be superfluous too. It is a question of faith and not of technique, it is an unbribable background, something that is imperiously and imperatively imposed on us, something that requires us to be faithful to ourselves and without which our life would be a failure, something like a "primordial emotion of duty" ([15]: 33).

With vocation we are not only asking ourselves about the authenticity of the one who exercises a profession, but also about responsibility. True vocation has nothing to do with regulations or laws:

> Professional ethics springs, like a spontaneous flower, from vocation. When the teacher discovers in the pupil a true vocation and encourages it, and when he sows knowledge in the field of demonstrated vocation, he is making not only a good doctor, but a good doctor with a profound professional morality. Hence my somewhat revolutionary conviction that there is no need for express rules of morality or courses of Deontology. In medical schools, morality, as a subject, is not usually taught. And this, which scandalizes some, has this fundamental reason. The doctor who is well prepared in the human and integral sense that we have explained, the doctor of vocation and not of pure technique, does not need regulations for his rectitude. For the ill-prepared physician, rules and moral advice will be perfectly useless. Here, as in all problems of moral conduct, laws are superfluous. ([15]: 348)

2.3 Virtue and Character in the Performance Society

When in an occupational activity we ask the question of vocation, there are underlying ethical questions that sometimes go unnoticed in the sociological or legal study of professions. This search for authenticity and fidelity to oneself is not easy in today's work ethics, especially because questions such as the search for excellence, the adequacy between vocation and profession, stability in the workplace, and the reconciliation with family life are problems without which a EP cannot be considered today.

In his book *After Virtue*, Alasdair MacIntyre raises an interesting reflection on the relationship between the activities, roles or jobs we perform and the whole of human life. He proposes a *narrative reconstruction of identity* that can be very useful for analyzing the relationship between professional life and the other dimensions of our life. It calls for a narrative conception of virtue so that the goods that we seek in what we do can make sense. This is important for a EP because "I can only answer the question of *what I am going to do* if I can answer the previous question of what story or stories I find myself part of" ([14]: 206).

This gives rise to a narrative conception of the profession because we are not only *agents* who carry out an occupational activity or *actors* who fulfil certain roles, but we are *authors* who construct reality. MacIntyre also speaks of *co-authors* because, in a narrative conception of personal or professional identity, each person is that for which he or she is justifiably held by others. In addition to being "accountable" to others, I am also someone who "can ask for accountability":

> ...I am not just someone who is accountable; I am someone who can always call others to account, who can call others into question. I am as much a part of their stories as they are a part of mine ... this asking and giving of accounts plays an important role in the constitution of narratives. To ask you what you did and why, to say what I did and why, to weigh the differences between your statement of what you did and mine of what I did, and vice versa, are essential constituents of all narrative .... Thus, without the responsibility-explainability of the self, the series of events that constitute all narrative could not occur ... without this responsibility-explainability, narratives would lack the continuity that makes both them and the actions that constitute them intelligible. ([14]: 269)

As we can see, this narrative approach to EP situates the *horizon of responsibility* in a humanistic and integrating sphere, because it limits work ethics to the world of duties, rules, and obligations of the profession itself. In this way, a single horizon of moral virtues appears which not only analyzes daily life from the perspective of work, but also analyzes work from the perspective of daily life as a whole. Two important challenges arise for the articulation of professional identity with personal identity.

On the one hand, there is the difficulty of articulating a certain personality or character over time, because the jobs, work, and occupations that are offered to us are of *short duration.* The new global capitalism and the "institutions" or companies that shape it have a restricted historical consciousness limited to the short term. Traditions, virtues, and values are narrative categories that require a solid historical consciousness that is not limited to the immediacy of the present, that is, a horizon of responsibility that includes the long term. A horizon that opens up when we consider sustainability or responsibility towards future generations, which, for the moment, is absent in the occupational activities of the new capitalism. In the words of Richard Sennett:

> Economic leaders and specialized journalists emphasize the global market and the use of new technologies, two aspects they see as the hallmarks of the capitalism of our time. While this is true enough, they do not contemplate another dimension of change: new ways of organizing time, and especially working time. The most tangible sign of that change might be the slogan: "nothing for the long term." In the realm of work, the traditional career that advances step by step along the corridors of one or two institutions is weakening .... Today, a young American with at least two years of college can expect to change jobs at least eleven times in the course of his or her working life, and to change his or her skill base at least three times during the forty years of work. ([17]: 20)

After analyzing the stories of numerous educated employees condemned to wander in scattered jobs and longing for a minimum of professional stability, Sennett describes the situation in dramatic terms.

> ... how to protect family relationships from succumbing to the short-term behaviors, the immediate mindset, and, basically, the weak degree of loyalty and commitment that characterize the modern workplace. Instead of the shifting values of the new economy, the family should value obligation, honesty, commitment, purpose .... How can a human being develop a narrative of identity and life history in a society composed of episodes and fragments ... short term capitalism threatens to corrode the character of workers, especially those aspects of character that bind human beings together and give each a sense of a sustainable self. ([17]: 25)

On the other hand, the performance imperative as a new way of approaching "professional duty," discipline, obligations, and rules. Ethical problems related to obedience or disobedience to work rules have been replaced by the performance imperative. Productivity, calculation, and measurement of performance not only condition "jobs" but are coming to condition all professions. Companies, hospitals, universities, and public administrations have fallen into the trap of a culture of performance that tends to confuse professional excellence with performance, calculation, and utility. They have been able to show this change from a disciplinary society to a performance society:

> ... Foucault's disciplinary society, consisting of hospitals, psychiatric hospitals, prisons, barracks, and factories, no longer corresponds to today's society. In its place a completely different one has long since been established, namely: a society of gyms, office towers, banks, airplanes, large shopping malls, and genetic laboratories. The society of the 21st century is no longer disciplinary, but a performance society. Its inhabitants are no longer called "subjects of obedience," but "subjects of performance." These subjects are entrepreneurs of themselves. Those walls of disciplinary institutions, which delimit the space between the normal and the abnormal, have an archaic effect .... The disciplinary society is still governed by the *no*. Its negativity generates madmen and criminals. The performance society, on the contrary, produces depressives and failures .... The syndrome of occupational burnout does not reveal an exhausted self, but rather an exhausted, burnt-out soul .... In reality, what is sickening is not the excess of responsibility and initiative, but the imperative of performance, as a new mandate of the late-modern work society. ([10]: 25–29)

3 Professional Excellence and Accountability in an Institutional Framework

3.1 Professional Life as Responsible Training

When we speak of *professional ethos* we refer to a way of doing things that forges a certain identity. We describe the set of *practices* of a given profession, but what do we mean by the term "practices"? There will be those who will contrast it with "theory," as if practices did not require knowledge and know-how. We have already seen, in describing a profession, that there is a productive circularity between knowing and doing, between theory and praxis, between knowledge and action. In this sense, EP does not turn its back on theory or on the foundations of practice, but demands it on a permanent basis. In the *professional ethos,* problems, concerns, and questions arise, so that it is always necessary to update knowledge.

In order to clarify our use of the concept of practices in EP, we can look at the anthropological structure of human activities. If we analyze in detail the relationship between what we do and who we are, we discover that there is an important process that we can learn about:

(a) Our will is expressed in the world through our *actions*. Acts are characterized by their voluntariness and transience.
(b) When an act lasts in time, it is called an *action*. The action is not fleeting but persistent and, although it is not always orderly or logical, it has claims of duration.
(c) When acts are carried out with a certain intentionality and we pursue a certain goal, then we speak of *actions*. Although in everyday language action and performance are synonymous, in EP we associate actions with purpose, meaning and intellection. Actions express intentional acts. It is a human act which, in addition to its duration, has intention.
(d) There are situations, contexts, and circumstances in which we do not know how to act, that is, moments in which our intelligence does not have a single answer to guide us in how to act. In these cases the professional is not only an agent or actor, as we have seen before, but becomes an "author" because he/she has to take risks, make decisions and get involved. In these cases, the professional is presented with a series of options or possibilities, and is forced to choose. This forcing of the decision to act is described as a process of appropriation of possibilities. Professional training is a process of appropriation of possibilities in which, when we decide or act, we train ourselves in one direction or another. With this process professionals become empowered, which means that reality gives them a power, that they are in reality with a certain power and that this power is real. So we are not talking about acts or performances, but about *activities or practices*.

Although practices take into account the past (actions), they are always oriented towards the future, because we endow them with intelligibility, reasonability, and explicability. Activities not only produce a change in the world, but also initiate processes and trends. Not only the external reality is transformed, but also the inner reality, that is to say, the identity in which we train ourselves. For example, in professions such as medicine, law, or education, when we appropriate some possibilities and leave others, we are directly intervening in the external and internal reality [20]. Although this process is complex, it is at the basis of a EP that seeks the continuous improvement of what we do and who we are. Professional life is not limited to a simple process of technical or instrumental training, but also involves reflective, deliberative and accountable training. That is why we call it "responsible training" [4, 6].

3.2 From Action to Practices: Internal Goods and External Goods

In chapter 14 of *After Virtue*, MacIntyre historically reconstructs the concept of practice and defines it as follows:

> By practice we mean any coherent and complex form of cooperative human activity, socially established, through which the goods internal to it are realized while attempting to achieve the models of excellence that are appropriate to that form of activity and partially define it, with the result that the human capacity to achieve excellence and the human concepts of the ends and goods involved are systematically extended. ([14]: 233)

If we apply this definition to some professions that are closer to us, such as doctors, lawyers or educators, we discover the following:

- These activities are not only individual but also social, and require cooperation (they are carried out together with other professionals, in an institutional framework);
- In those that "realize" a series of internal goods (a series of fruits are produced in their realization, the professional has self-esteem, self-realization);
- There are models of excellence that partially define the profession (there is a teacher or model that serves as an example);
- Professional capacity (in particular) does not stand apart from human capacity (in general), where there are also proper ends and goods (there is some element related to vocation),
- There is a systematic relationship between professional training and human training (in addition to exercising a profession, we exercise as "persons").

MacIntyre does not refer to these professions in particular, but to the game of chess as a paradigm or model, and from this he delimits what will be internal goods and external goods:

> There are two types of goods that are gained by playing chess. On the one hand, external and contingent goods attached to chess or other practices because of other social circumstances... such as prestige, rank, or money. There are always alternative ways to achieve these goods, which are not obtained just by engaging in some particular kind of practice. On the other hand, there are internal goods that cannot be obtained except by playing chess... We say that they are internal for two reasons: first, because they are only realized in that game; second, because they can only be identified and recognized by engaging in that particular practice. Those who lack relevant experience cannot judge about these internal goods. ([14]: 235)

In this context, all practice involves goods, models of exemplarity or excellence, and obedience to rules. Entering a profession and participating in a *professional ethos* requires accepting the authority of these models, and the limitation of one's own perspective. I participate in the practices of a profession when I align my attitudes, preferences, and tastes with the models that define the practice. When I start working in a hospital, an administration, or a school there are previous models, and that is why "we cannot start in a practice without accepting the authority of the best models made so far" ([14]: 236). In this sense, as the hermeneutic ethics reminds us, the rehabilitation of authority is an important element when we raise the practical reason in the horizon of responsibility [2].

There is an important difference between internal and external goods. External goods are always the possession and property of an individual; they are also the object of competition in which there are losers and winners. Internal goods, however, are the result of competing in excellence, and their realization, even if personal, is a good for the whole community that participates in that practice.

3.3 From Practices to Institutions and Virtues

The concept of practice implies that its goods can only be achieved by subordinating ourselves, within the practice, to our relationship with others. When we share the models and purposes of practices, we establish a mutual relationship, whether we want to or not, with reference to shared models of justice and value. What distinguishes a practice is the way in which the relevant goods and purposes served by technical skill are transformed and enriched by that extension of human faculties, and in consideration of those internal goods that define each type of practice ([14]: 237, 240). Each practice has its own history, demands its own learning, and along with the technical dimension has an ethical dimension.

Following MacIntyre, we should note that practices should not be confused with institutions. No practice can survive for long if it is not sustained by institutions:

> … so intimate is the relationship between practices and institutions, and consequently that of internal goods to external goods, that institutions and practices typically form a single causal order, where the ideals and creativity of the practice are always vulnerable to the greed of the institution, where the cooperative attention to the common good of the practice is always vulnerable to the competitiveness of the institution. In this context, the essential role of the virtues is clear. Without them, without justice, courage, and truthfulness, practices could not resist the corrupting power of institutions. ([14]: 241)

What MacIntyre calls "corrupting power" has to be properly understood and therefore he qualifies that it refers to the need to:

> form and sustain forms of human community, and therefore institutions, which have all the characteristics of a practice …. The exercise of the virtues itself is likely to demand a very definite attitude to social and political questions; and we learn or fail to learn the exercise of the virtues always within a particular community with its own specific institutional forms. ([14]: 242)

There are two models of institutional ethics that MacIntyre presents. On the one hand, there is the approach of liberal individualism where the community is only the playing field where each individual pursues the ideal of the good life that he or she has chosen and, therefore, political institutions only exist to guarantee order. In this case, political institutions cannot impose any ideal of life, and must remain neutral. On the other hand, there is the communitarian or neo-Aristotelian approach, according to which the political community not only demands the exercise of virtues for its own maintenance, but one of the obligations of parental authority is "to force children to become virtuous adults" ([14]: 242).

Although these two models require nuances and it is difficult for them to be pure, it is important to point out the role of virtues as mediators, which dynamize the relationship between practices and institutions. When we ask ourselves about the virtues of a given profession, we are not only analyzing the relationship between practices and institutions, but also the relationship between internal and external goods. Here MacIntyre's affirmation is important when he points out that virtues do not maintain the same relation with internal goods as with external ones. In his view, the possession of the virtues is necessary for the attainment of internal goods. It is also true that the possession of the virtues may very well prevent us from achieving external goods: "the cultivation of truthfulness, justice and courage, the world being contingently as it is, will often prevent us from being rich, famous and powerful" ([14]: 243).

Following these arguments, MacIntyre provides us with an interesting approach to virtue:

> The virtues are to be understood as those dispositions which not only sustain practices and enable us to achieve the goods internal to the practices, but which will also sustain us in the relevant kind of pursuit of the good, helping us to overcome the risks, dangers, temptations and distractions we encounter, and providing us with increasing self-knowledge and increasing knowledge of the good. ([14]: 270)

The process of professional training to which we have referred, and the recovery of this integrating and mediating function in the full sense of the word, leads us to the concept of tradition within a certain *professional ethos*. As in Gadamer's hermeneutics, we are faced with a narrative, reflexive, and critical recovery of the value of tradition:

> What maintains and makes traditions strong? What weakens and destroys them? Fundamentally, the answer is: the exercise of the relevant virtues or their absence. The virtues find their end and purpose not only in maintaining the relationships necessary for the multiplicity of goods internal to practices to be achieved, and not only in sustaining the individual way of life in which the individual can seek his or her good as the good of the whole of life, but also in maintaining those traditions that provide both practices and individual lives with their necessary historical context. Lack of justice, lack of truthfulness, lack of courage, lack of the relevant intellectual virtues, corrupt traditions as well as the institutions and practices that derive their life from those traditions, of which they are contemporary incarnations. To admit this is also to admit the existence of an additional virtue, the importance of which is all the more obvious the less present it is: that of a proper sense of the traditions to which one belongs and with which one is confronted. This virtue is not to be confused with any form of conservative antiquarian mania .... It is rather that the proper sense of tradition manifests itself in an understanding of the future possibilities that the past makes available

> to the present. Living traditions, precisely because they are an as yet uncompleted narrative, confront us with the future, whose determinate and determinable character, insofar as it possesses it, derives from the past. ([14]: 274–275)

4 Ethics and Codes of Ethics

4.1 Between Credibility, Legitimacy, and Excellence

From a historical point of view, the different professionals have been organized into groups, guilds, unions, or professional associations. As professions have evolved socially, their virtues and traditions have also adapted. This has led to the creation of guilds of workers and professionals who play important roles in social life. When these groups of workers or professionals are concerned only with their own interests as labor collectives and are not open to other activities, other jobs, and other needs of society, then *corporatism occurs*. How do we prevent groups of workers or professionals from closing in on themselves, but instead promoting the *common good*, *social justice* and, in general, the values of an *open society*? Is there any possibility that the virtues and traditions of a given labor collective are not at the service of *particular* interests, but of *universal* interests?

Although today we have numerous resources available to answer these questions and analyze the ethical dimension of groups of workers or professional collectives, the most common resource in the history of ethics has been professional codes. Although they are used as a resource to analyze, evaluate, or measure the way in which a company, organization, or group is "accountable" to society, codes of ethics have their origins in the practices, virtues, and traditions of a particular profession that acts as a paradigm. In fact, when we review the history of codes we always refer to the Hippocratic oath and the code of the medical profession [9].

Codes can be PD tools and can be used to train the professionals who make up this collective, as if professional expertise were regulated and codified. However, deontological codes are useful tools in EP because they contribute to *dynamizing the work collective so that it has cultural credibility, social legitimacy, and professional excellence*. There are three important levels in this dynamization when we analyze the relationship between the individual and the organization, between the subject who is joining a profession, and the collective that has historically regulated it. There is a level in which the individual is unaware of the conventions, norms, or codes of a profession (preconventional level), another in which he/she aligns or adjusts to them (conventional level), and another in which he/she participates directly in their dynamization, change, or modification.

It is important to approach deontological codes from this dynamic process, and EP plays a role in this. Without the ethical perspective, deontology loses its horizon of reference; without the motivations for action, protocols and standards have only a mechanical function. A deontological code is a set of duties or rules that regulate a *professional ethos*, i.e. access to the profession, the relationship between professionals, conflicts, training, control and, above all, the levels of credibility, legitimacy, or excellence of an occupational activity. The code is a basic tool for what we would call quality assessment *of* a profession and *in* a profession, i.e. the quality of professionals in the eyes of society, and the quality of professionals in the eyes of their own colleagues or peers.

Before analyzing the advantages and disadvantages of having a code of ethics and the corresponding deontological commission of the professional group or association where the professionals are registered or accredited, it is important to take a look at this text by Professor Hortal that can help us in the elaboration or application of the codes:

> In order to correctly interpret a code of ethics it is necessary to take into account the situation of the corresponding profession, whether it is in a phase of ascent and growing social recognition or whether it is in crisis, whether it is already consolidated and stable, cohesive or divided, what type of internal leadership and social recognition it has, what scandals there have been or there are, what support the writers of the code have, whether it is a code to write or keep or a code that keeps alive the debate on the profession and the best way to regulate it. ([11]: 197)

4.2 Advantages and Disadvantages of Codes

As Lozano ([13]: 230–231) has pointed out, the history of ethical and deontological codes has gone from the simple regulation of conflicts of interest to the regulation of environmental sustainability issues. In general, codes of ethics regulate in writing the shared responsibilities of a professional group, and publicly express the criteria, goals, and values that identify them. Although public administrations, companies, and organizations use codes of ethics as tools for legitimization, credibility, and social plausibility, we are only going to describe the double face of the codes of professional associations. In open societies where professions are undergoing rapid change, EP reminds us that there are advantages and disadvantages to having a code of ethics.

These are the disadvantages of codes of ethics:

(1) They express and objectify the degrees of commitment and motivation, but do not generate them. They take motivation, illusion, and enthusiasm for professional life for granted.
(2) They favor normativism with the risk of hiding and concealing personal responsibilities in organizational rules.
(3) They do not in themselves generate a capacity for discernment and judgement needed when applying what is stipulated in them.
(4) The agreement to which they respond favors organizational consensus, but also imprecision, generalities and, organizational vagueness.
(5) They can become a refuge for evading responsibility, and justifying practices that are not accepted in any other profession, morally unclear, or socially dubious.

(6) They run the risk of becoming cosmetic instruments of moral *marketing*, i.e., public relations for the organization. There is the risk of turning ethics into cosmetics.
(7) They imply an organizational or corporate compartmentalization of society and jobs, especially when they leave aside interaction with society. There may be cases where the corporate justice of professionals does not coincide with the social justice of the people.
(8) They reduce ethical issues to only those that are covered and recorded in writing in the code.
(9) The articles of the codes are identified and confused with administrative and legal norms. In addition to this confusion with the rules contained in other codes external to the profession (civil code), they can also be confused with the internal regulations of a given group, confusing *professional ethos* with administrative or organizational procedures.
(10) Codes are instruments for organizational control and not for cooperative motivation. Rather than being an invitation to work together in a profession, they establish rules and control in the profession.

On the other hand, these are the advantages of a code of ethics:

(1) They avoid opportunism and subjectivism in the search for professional performance criteria.
(2) They clarify the expectations and references of those who are directly affected by the practices of a given profession. This clarification facilitates the knowledge of minimum standards that are required of everyone in the group.
(3) They formalize shared references and facilitate conflict resolution. In this way they help to clarify the relationship between internal and external goods.
(4) They aim at a minimum moral commitment of a public nature with which to resist the possible temptations of transgression, anomie, voluntarism, or misunderstood heroism.
(5) They hinder arbitrariness in the processes of interpretation of situations, circumstances, and decisions.
(6) They catalyze the changes and transformations that can occur in a profession over time. Codes are very useful when a professional group wants to respond effectively to social changes because they help it to organize experiences and expectations.
(7) They facilitate communication, training, and the generation of a shared organizational culture.
(8) They recall the moral dimension of the profession because they provide interpretative keys to jointly think about the technical, anthropological, and ethical dimension of human activities.
(9) They offer minimum requirements for all professionals who carry out a series of common activities, who have common aspirations, and who make common demands on themselves.

(10) They institutionalize processes of social and community pedagogy because they facilitate the processes with which a profession is associated over time. For example, they are useful tools in the training, integration, application of cases, and critical adaptation of new professionals who, periodically, access a certain group.

4.3 Demarcation, Conflicts, and New Professions

When a collective decides to have a code of ethics, it is doing something important in two ways: internally, because the team of professionals who agree on it establishes self-imposed rules to achieve excellence in the practices that give meaning to their profession; and externally, because in an orderly and systematic way it offers society the values, rules of the game, and goals of their profession. Moreover, it strengthens an important public ethical space which, on the one hand, is not strictly legal or enforceable by positive law and, on the other hand, is not strictly individual or subject to the arbitrariness of the workers. Codes of ethics strengthen civil society and generate social capital; they nourish this area of public ethics in which a political community or a professional interested in the trust that he or she needs from his or her clients can develop.

The history of some codes of ethics, such as those of medicine, law, or education, shows that codes are necessary but not sufficient to maintain the legitimacy, credibility, and plausibility of a profession. Although codes regulate the relationships between professionals, there are currently multiple levels and very different relationships with other groups within the same profession. They cannot regulate all the forms of collaboration, coordination, and hierarchy that, in fact, occur in professional life. We live in complex organizations and plural societies where not all professionals have the same job stability, administrative level, or leadership. This makes it necessary to consider how a EP should approach internal conflicts among its own professionals.

One of the most important challenges facing EP today comes from the diversity of new activities and practices that are emerging in the information society. New occupations, new jobs, and novel jobs which may or may not end up as professions are emerging. I would like to give two easily recognizable examples. First, the field of Computer Engineers brings together not only those who trained and qualified in Computer Science, but a whole group of professionals in Mathematics, Physics, Documentation, or Electronics who carry out practices that are changing at an accelerated pace. Second, the field of Property Administrators, a socially very important group because it provides services to communities of owners, but there is no university degree, a career, or a standardized package of knowledge that is required to enter it. We are facing a new profession with a long history, whose members are taking on new services and functions in the communities of owners.

If in these two professional groups there are structural problems in establishing a clear demarcation between one practice and another, remember the numerous professions that are emerging in the social and healthcare field. The different professional specialties and the new social demands pose a major challenge for EP. Between a work ethic where the employee's activity is legally regulated by labor law, and a public administration employee ethic where the activity is bureaucratized or almost entirely protocolized, EP is called upon to play new and important roles in society. This is especially the case when, in the development of talents, the training of professionals, and the continued service to society, we are looking for a certain stability or a certain long-term perspective that allows us, at least, to avoid the corrosion of character.

Commenting on the consequences of the new models of capitalist and post-industrial societies, Richard Sennett recounts the changes in values that occur in workers, and the need to measure time differently:

> … you used to be a liberal … now you've become a cultural conservative … you want to maintain social relationships and offer lasting guidance. It is against the broken bonds at work, against the deliberate amnesia of your neighbors and the specter of your children becoming "mall rats" that you postulate the idea of enduring values … the special characteristics of time in neo-capitalism have created a conflict between character and experience, the experience of a disjointed time that threatens people's ability to consolidate their character into enduring narratives …. What is unique about uncertainty today is that it exists without the threat of historical disaster; instead, it is embedded in the everyday practices of a vigorous capitalism …. The corrosion of character may be an inevitable consequence. The slogan "nothing in the long run" disorients planned action, dissolves the bonds of trust and commitment, and separates will from behavior…. ([17]: 27–31)

5 Conclusion: Trust and Social Responsibility

As we have seen, the EP have made their way in open, new and complex societies. They have not broken their ties with the PD, nor have they turned their backs on the deontological dimension of the professions. However, the historical, ethical, and anthropological reconstruction of the profession as an activity can be of great help in gaining a better understanding of the complex world of *institutionalized human action*. Today, it is difficult to propose a EP without referring to the precariousness of work, the instability of jobs, and the difficulties that groups or groups of workers have in order to be considered "professionals."

EP has to be approached from a more ambitious horizon than that of an ethic of employees, workers, or civil servants. This is why we have focused on the historical complexity of the concept of profession, developing it in three dimensions. First, as a touchstone for a *strong civil society* trying to make its way through an economy that lacks sensitivity for the long term, and a politics that tries to patrimonialize social activities. Secondly, as a tool to *energize the complex world of virtues, values, and public ethics* when we are witnessing a seemingly unstoppable trend towards moral individualism. Thirdly, as an *opportunity to promote healthy, mature, and balanced personalities* who are able to articulate work and family, business and leisure, activity and reflection, profession, and vocation.

These dimensions of professional life are kept alive when we do not exclude the question of the *vocational dimension within professional ethics*. However difficult it may be to consider a job, an occupation, or one of the new professions in "vocational" terms, at some point we will have to consider the relationship between what we do and who we are. In conclusion, I leave here an interesting text by N. Bellah, with which to continue promoting EP:

> The demand to become someone through work is a requirement that ... encompasses different notions of work and how it influences the way we are. In the sense of a "job," work is a way of earning money and supporting oneself. It maintains a self [that is] defined by success and financial security, and by all that money can buy. In the sense of a "career," work marks one's advancement through life for success and promotion in an occupation. It produces a self [that is] defined by a broader kind of success, including status and social prestige, and by a sense of increasing power and competence that make work itself a source of self-esteem. In the most important sense of vocation, work constitutes a practical ideal of activity and character that makes a person's work morally inseparable from his or her life. It includes the self in a community of disciplined practice and sound judgment whose activity has meaning and value in itself, and not merely for the product or profit that results from it. But vocation not only binds a person to his or her fellow workers. It also establishes a bond with the community in general, a whole in which the vocation of each one contributes to the good of the whole .... The idea of vocation is more and more difficult to understand ... with the appearance of the industrial society on a large scale it was more difficult to see work as a contribution to the whole and easier to interpret it as a fragmentary and selfish activity ... in spite of being replaced by those of employment or professional career, something of this idea of vocation continues existing, not necessarily in opposition, but [in] complement of employment and career .... ([1]: 96 ff.)

References

1. Bellah N et al (1989) Habits of the heart. Alianza, Madrid
2. Domingo A (1989) The art of being able to not be right. The dialogical hermeneutics of Gadamer. Ediciones de la Universidad Pontificia, Salamanca
3. Domingo A (1994) Ethics of management. In: Cortina A, Conill J, García-Marzá D, Domingo A (eds) Ética de la empresa. Trotta, Madrid, pp. 95–123
4. Domingo A (2005) Ethics of the professions and university education: three models of professional responsibility. Revista de Fomento Social 60:39–55
5. Domingo A (2009) The disenchantment of applied ethics. Daring to think nothing less than an ethics 'without more'. In: Hortal A (ed) Ética pensada y compartida. Homage to Augusto Hortal. Ediciones Univ. Pontificia Comillas, Madrid, pp 51–72
6. Domingo A (2010) Ethics of the professions and university education: three models of professional responsibility. In: Correa M, Martínez P (eds) La riqueza ética de las profesiones. RIL editors, Santiago de Chile, pp. 91–109
7. Domingo A (2014) The art of caring. Atender, dialogar y responder. Rialp, Madrid

8. Fernández JL, Hortal, A. (comp.) (1994) Ética de las profesiones. Ediciones de la Universidad Pontificia Comillas, Madrid
9. Gracia D (1989) Fundamentals of bioethics. Ediciones Universidad Complutense, Madrid
10. Han B-C (2012) La sociedad del cansancio. Herder, Barcelona
11. Hortal A (2002) Ética general de las profesiones. Desclée de Brower-U. P. Comillas, Madrid
12. Hortal A, Etxeberria X (2011) Professionals and public life. Desclée de Brower-U. P. Comillas, Madrid
13. Lozano JF (2007) Ethical codes and ethical audits. Veritas 17:225–251
14. MacIntyre A (1987) After virtue. Crítica, Barcelona
15. Marañón G (1973) Obras completas, vol IX. Espasa Calpe, Madrid
16. Ricoeur P (2008) Studies, readings and exercises in applied ethics. Trotta, Madrid
17. Sennett R (2000) The corrosion of character. The personal consequences of work in the new capitalism. Anagrama, Barcelona
18. Weber M (1981) The politician and the scientist. Alianza, Madrid
19. Weber M (2001) La ética protestante y el espíritu del capitalismo. Alianza, Madrid
20. Zubiri X (1989) Estructura dinámica de la realidad. Alianza-Fundación ZX, Madrid

Neuroethics

7

Adela Cortina

Abstract

Neuroethics was officially born at the beginning of the twenty-first century, thanks to the advance of neuroscience, as the ethics of neuroscience and the neuroscience of ethics. The former attempts to develop an ethical framework for regulating behavior in neuroscientific research and its application to humans, while the latter refers to the impact of neuroscientific knowledge on our understanding of ethics itself. This chapter aims to briefly discuss neuroethics in its dual aspect and to address the challenge of naturalism in neuroethics.

Keywords

Applied ethics • Neuroscience ethics • Fundamental neuroethics • Naturalism

1 A New Knowledge

Neuroethics was officially born in 2002 at a conference held in San Francisco on May 13 and 14, and sponsored by the DANA Foundation. Apparently, it is Pontius who first used the term "neuroethics" in 1973, and later Cranford [27] and

This study is part of the research project "Ethics and Democracy in the face of the challenges of Artificial Intelligence" PID2019-109078RB-C22 funded by MCIN/AEI/https://doi.org/10.13039/501100011033 and the activities of the research group of excellence PROMETEO/2009/085 of the Generalitat Valenciana.

A. Cortina (✉)
Universidad de Valencia, Valencia, Spain
e-mail: adela.cortina@uv.es

A. Cortina and M. Correa Casanova (eds.), *Applied Ethics from Medicine to Humor*, Integrated Science 21, https://doi.org/10.1007/978-3-031-59009-2_7

Churchland [24] ([1]: 96–98, [6]: 50–52).[1] To the latter also belongs the merit of having published *Neurophilosophy: Toward a Unified Science of the Mind-Brain* in 1986, a pioneering, extremely complex, and problematic book.

The birth of neuroethics is based on the extraordinary boom experienced by neurosciences, especially from the 1990s, a decade that the United States Congress declared as the "Decade of the Brain." In 2013, President Barack Obama presented the *Brain Research through Advancing Innovative Neurotechnologies* (BRAIN) project. Also in 2013, the European Union established *The Human Brain Project* (BP), coordinated by neuroscientist Henry Makram.

Naturally, there are many ways to characterize neurosciences; put simply, they are experimental sciences that, with all the technical tools available and using the scientific method of observation, experimentation, and hypothesis, tend to explain how the brain works ([65]: 35).

Undoubtedly, neurosciences have a long history, which starts with Hippocrates and has as inescapable milestones those represented by Thomas Willis' treatise *Cerebri Anatome* (1664), John Harlow's article, narrating the famous accident suffered by Phineas Gage, later collected by Antonio Damasio in *Descartes' error*, and, of course, the research of Ramón and Cajal ([1]: 23–27, [5]). But since the end of the twentieth century, a series of advances has brought neurosciences to the forefront of attention, especially progress in neuroimaging techniques.

With the help of these neuroimaging techniques it is possible to establish some correlations, for example, between the formulation of moral judgments and the activity of certain areas of the brain, bringing to light, albeit with great caution, the neural basis of human activities, including morality. "With great caution" because the reading of brain images, which seemed to ensure an enormous advance in knowing which areas of the brain are activated when certain behaviors are set in motion, is now showing enormous limitations. We do not know well what "correlates" mean and, moreover, a great deal of interpretation is needed to draw conclusions from brain images [17, 19, 32, 55].

In any case, in recent times the prefix "neuro" precedes the name of a large number of fields of knowledge, such as neuroeconomics, neuropolitics, neuroreligion and a long list of other "neuros." The "neuro" has become transversal, because knowing the neural bases of any of these human activities, even with all the difficulties, helps to understand them better, which is a great advance [19].

However, the fact that neuroscientific research involves a large amount of money has led to suspicions that it is above all a profitable business, that it promises more than it can deliver, and that it will be applied not only to therapy, but also to corporate or political *marketing* for manipulative purposes, and that neuroscience may even become a new ideology. A critical neuroscience is therefore emerging [22, 23].

[1] Gazzaniga publishes a first recognized monograph on neuroethics in 2005 [35], Judy Illes publishes the first important collective on the subject in 2006, and later Walter Glannon in [39]. For the birth of neuroethics, see also, among others: Illes [50], Giménez and Sánchez-Migallón [38] and Bonete [6].

It is in this context of neuroscientific fervor as well as caution, suspicion, and criticism, that neuroethics was born. According to some of the papers presented at the San Francisco Congress, neuroethics is "the study of the ethical issues raised by brain research and the application of discoveries to medical practice, legal interpretations, and health and social policy" ([63]: III), or "the examination of right and wrong, good and bad, in the treatment of the human brain, in its enhancement, or in the undesirable invasion of the brain or its troubling manipulation" ([63]: 5).

Certainly, if neuroethics were nothing more than this, it would be an applied ethics and would follow the principles born of the Belmont Report, or else it would understand that non-maleficence, beneficence, autonomy and justice constitute the common language with which these issues should be dealt ([54]: 423–425). In fact, neurosurgeon Henry Marsh's book *Do No Harm* has become a bestseller. The English translation "Above all, do no harm" perhaps does not do justice to the force of the Hippocratic oath and especially to the negative injunction "thou shalt not harm" which, as a duty of perfect obligation, is expressed in the principle of non-maleficence. In any case, the book aims to expose the difficulties of fulfilling this duty by making decisions about actions in an organ as complex as the brain; and certainly, neuroethics thus understood is a promising applied ethics. But there is another meaning of neuroethics that makes it a discipline concerned with the bases and foundations of morality.

Accepting a famous suggestion by Roskies, it is possible to distinguish two branches of neuroethics: (1) The ethics of neuroscience, which attempts to develop an ethical framework for regulating behavior in neuroscientific research and in the application of neuroscientific knowledge to human beings; (2) The neuroscience of ethics, which concerns the impact of neuroscientific knowledge on our understanding of ethics itself, is concerned with the neural basis of moral agency ([56, 72]: 1–2).

This second branch would consist of an interdisciplinary knowledge that claims to be fundamental ethics, and that can come to question ethical theories that are not compatible with the discovered neural bases ([18, 56]: IX).

2 Neuroethics as Applied Ethics

The ethics of neuroscience analyzes the ethical correctness or incorrectness of neuroscientific research and its application, using the language of non-maleficence, beneficence, autonomy and justice. Without seeking to be exhaustive, it faces specific problems, such as: whether brain readings can infringe on the right to privacy [13]; whether evidence taken from brain images can be admitted in court to exonerate the accused; whether it is legitimate to use new techniques in the field of national security, in relation to the "war on terror" and the like, opening the way to "neurosecurity" [66].

In all these cases it is important to treat with care the data obtained through the alleged "brain readings" so as not to: stigmatize certain people; violate the duty of confidentiality by using the knowledge for police purposes, in the fight against terrorism, for labor or eugenic purposes; and, or to exempt from responsibility to those who could act otherwise. And in all these cases it is essential not to generate expectations that cannot be fulfilled, or generate hopes that are not based on rigorous studies [17, 19].

The ethics of neuroscience is also faced with the possibilities of more adequately specifying problems such as the vegetative state, consciousness or brain death, as can be seen in Enrique Bonete's text *Practical Neuroethics*, which devotes the second part of the book to it and includes a very extensive bibliography on the subject [6].

Obviously, both the mentioned research and its possible applications raise ethical questions, but one of them has taken special prominence: the ethical legitimacy of *neuroenhancement*, that is, of *enhancement interventions*, and not only of therapy. Regarding this issue, "anti-enhancement" authors such as Michael Sandel, Leon Kass or Francis Fukuyama, "anti-anti-enhancement" authors, among which Julian Savulescu, Nicholas Agar, Dan Brock, Nick Bostrom, David DeGrazia, Anders Sandberg or Thomas Murria would be placed, and an intermediate position, represented very aptly by Buchanan ([7]: 1–33) and Cortina ([21]: ch. 6), confront each other.

Clearly, debates about the legitimacy of enhancement have public policy implications [73]. They also take us beyond neuroethics, understood as the ethics of neuroscience, to the neuroscience of ethics. To speak of "enhancing" normal capacities forces the questions of what is "normal" and what is "abnormal," and whether, in case what is normal can be determined, it should be considered as sacred. Is there a "human nature" that to be identified with normality, which must be regarded as untouchable? Is the method to access it that of the experimental sciences, or is it necessary to rely on philosophical methods? Is it a "naturalized" nature or is naturalism insufficient to account for it? [11–13, 37, 61, 62].

Undoubtedly, these questions take us beyond an ethics of research and the application of the advances made, which assumes the framework of already existing ethical theories, and directs us towards a *fundamental neuroethics*, which tries to determine what are the cerebral bases of moral conduct. Neuroethics, as the *neuroscience of ethics*, needs a different characterization because it is *fundamental ethics*, not applied ethics.

3 The Challenges of Fundamental Neuroethics

Fundamental neuroethics deals with problems similar to those that addressed by philosophy, but from a neuroscientific perspective, in the broad sense of the word. It draws on philosophy and data from neuroscience, genetics, experimental economics, psychology, anthropology, and evolutionary biology ([26]: 3, [75]: 33). From this approach, neuroethics has faced questions such as the ones we expose, among others [19].

The most frequent path to show that neuroethics constitutes an unique field is how challenging a good number of subjects have found the task of supporting their moral judgments with arguments. The subjects formulated an immediate, intuitive judgement, but when asked for reasons they were unable to provide any. Faced with this *moral dumbfounding*, authors such as Haidt and Hersch pointed out that our moral judgments are intuitive, associated with emotion, and only later do we try to support them with reasons [47, 48].

Taking this common thread, Greene and his collaborators wonder if the reason for this is that in our brain there are moral codes, coined through evolution, and associated with emotions in the process of brain formation, which would explain its intuitive and emotional character. The brain readings performed on subjects who were asked about their position in moral dilemmas seemed to support this hypothesis; the respondents answered more quickly when the dilemmas were personal than when they were impersonal, supporting the hypothesis of the strength of emotion in the former and the greater need for rational reflection in the latter. The evolutionary hypothesis that moral codes closely connected with emotion were coined in the brain during its formation, gained strength [41].

The role of emotions in the formulation of moral judgments has become a *topos* of neuroethics, a commonplace [28, 29], which has only been reinforced [45, 46]. Works such as those of Jorge Moll and his collaborators are a source of progress in this area [64].

In addition, some authors believed it was possible to find an ethics that, having a cerebral basis, would be universal and could overturn traditional theories [35, 65]. Neuroethics became, *velis nolis*, a fundamental ethics. Other authors more prudently understood that, rather than a universal ethics, neurosciences in a broad sense allow us to discover an innate moral structure from which human beings can speak different "moral languages" from different cultures [49, 56]. Obviously, criticisms of this position were not long in coming [4, 71], but, in any case, the articulation between morality as structure and morality as content was back on track [3, 76].

Certainly, the thread that we have mentioned leads us to another classic topic of ethics. From another angle, Benjamin Libet's experiments once again put on the table the secular determinism-freedom aporia ([17] : 177–213, [57–59]). Some authors approached it from a mechanistic conception of the brain that leads them to decree that freedom is nothing more than an illusion (Wegner, Roth, Wolf Singer, Rubia); while others confronted it from a conception of the brain as an organ with the capacity to take the initiative (Changeux, LeDoux, Evers). Hard determinists or incompatibilists, and also soft determinists or compatibilists confront libertarians (who can also be compatibilists or incompatibilists) [17]. Also, a thriving hypothesis is the conviction that we have two types of language, which have emerged in the process of evolution, the first person language and the third person language, and that both are indispensable for human life. "Epistemic dualism must have emerged from a learning process and been accredited in the cognitive confrontation of *homo sapiens* with a world full of risks" ([17] : 195, [43]: 174, [44]).

Also, there is an attempt to use cerebral processes to address questions such as what is the nature of consciousness ([2, 20, 67, 69], Cortina 21: ch. 5); to what extent can we speak of the identity of the self; is the mind reduced to the brain, because "we are our brain"; or, are they two different entities (i.e., mind and brain) and, in that case, what is their relationship.

It is not surprising in this context that Aristotelian virtue ethics is back on the scene. But it is even more interesting that the secular opposition between deontologists and teleologists should once again enter the fray and, above all, that neuroscientific discoveries are used to characterize the former as emotivists and the latter as rationalists. While deontologists have been accussed of being unrepentant rationalists, teleologists were praised for their ability to attend to emotions ([31]: 10, [40, 42]). This is an issue that has important implications for the way we understand the function of law [8].

To conclude this list of questions, there is the issue of finding outwhich mode of political organization would best fit the cerebral bases of behavior ([17]: 99–148, [36]). It is also important to understand how moral education should be implemented, taking into account how the brain is constituted and how it develops—a task addressed by moral and political neuroeducation [9]—and are also critical the questions about the social justice raised by inequality in brain development due to material and cultural poverty [60].

These are classic problems now seen in a new light, which is always a gain. No matter how little we know about an organ as mysterious as the brain, getting an insight into its inner workings and trying to comprehend it in order to know more about ourselves is always a benefit. In the end, that is what philosophy has done in its moments of greatest splendor.

At the bottom of all these questions, there are two transversal issues that deeply affect the way neuroethics is developed:

(1) What is understood by morality when we try to detect the cerebral bases of this type of human activity, and what is its specificity compared to other ways of solving problems? Some insist on affirming that there is no gulf between humans and other mammals, that chimpanzees and bonobos, among others, also have social instincts and are capable of a certain moral life. The challenge is to find out what we mean by "moral."

(2) To what methods does neuroethics resort? Are they only scientific (in which case naturalism is served) or also philosophical? In the second case, it is important to know which philosophical methods are employed and how are they articulated with the experimental ones. The discussion between naturalism and non-naturalism is on-going.

4 The Challenge of Naturalism

In the long-term, some neuroethicists suggest devising a proposal for neuroethics, from the design of an integrative philosophical ethics, that unifies the advances made by neurosciences and the remaining cognitive sciences (Churchland, Levy). This ethics would be the most appropriate for understanding the different sides of human morality, taking into account what we are learning about the brain.

In this sense, Churchland's way of proceeding is paradigmatic. From the outset she announced his intention not to resort to any supernatural instance nor to any "unrealistic and rarified" concept of reason, in a clear allusion to the Kantian proposal. Following Hume, it is about discovering the roots of morality in our empirically observable nature: in how we are, what we take care of, and what we care about. From that empirical description of how we behave, *we will not derive* an "ought" from an "is," but *we will infer* what ought to be done, bearing in mind that "inferring" is much broader than "deriving" ([26]: 6).

This fear of falling into supernatural or metaphysical proposals is a common denominator of most works in neuroethics, which leads to opting for naturalism. Darwin's *evolutionary theory* already offered a naturalistic explanation of the origin of morality in *The Origin of Man* (1871), which seems to endorse Hume's description. Genetics, for its part, extracted consequences for moral qualification from the innate predispositions to act in one way or another. Meanwhile, the advance of neurosciences has put back on the table the pretension of explaining morality without the need to resort to anything other than empirically described nature. Naturalistic explanations along the lines of Hume and Darwin have triumphed to the point of becoming the orthodox position. Habermas himself, undoubtedly closer to Kant than to Hume, defends a "soft naturalism" that is difficult to understand ([43]: 161, [44, 68]). Naturalist authors include Haidt, Greene, Gazzaniga, Mora, Joyce, Kitcher (among many others), and a very comprehensive proposal by Patricia S. Churchland.

However, it is one thing to say what one intends to do, and another to actually do it. It is really impossible to begin one's search about the origin of morality without previously having some notion of morality. A notion that has not been extracted from empirically observable data, but carved out both culturally and experientially. Given the limited space we have, we will take Churchland's search for the origin of moral values as a paradigmatic example of this way of proceeding..

In Churchland's opinion, empirical data support the fact that the basic valuations of human beings are pleasure and pain (as Hume said) and because the human brain is social, like that of other mammals, they are associated not only with their own care and well-being but also with that of others. The brains of social animals are organized in such a way that they feel pleasure in the exercise of social dispositions, such as the care of children, companions, and relatives—those with whom there is a bond. Neurochemical substances such as vasopressin and oxytocin mediate the bonding of partners, parents and children, and probably of kith and kin. Other neurotransmitters, such as serotonin and dopamine, also play a role in social life, as do hormones such as testosterone ([25]: 103, [30, 52]). In fact, Churchland suggests that perhaps oxytocin is the basis of what Hume called "moral sentiment."

The likelihood that oxytocin, vasopressin, and a spectrum of receptors are important in human styles of sociality and their neurological understanding would then have broad implications for the origin and basis of human morality. Our brains would be structured to defend our interests and those of relatives and friends.

The question about how morality arises from this requires an evolutionary response. Human beings have social instincts, enabled by genes, and adjust them to the practices of their social environment by the reward system. These instincts are precisely the platform for cooperation and the maintenance of social order, and provide the biological foundation for ethics in its broadest sense. They provide the basis for attachment to mates, offspring and relatives, and friends.

Morality would then be a dimension of social behavior closely connected with our natural disposition to care for others. It would be a system of social behavior with four dimensions, made up of intertwined neural processes: (1) caring, rooted in the bond with relatives and friends, and care for their well-being; (2) recognition of others' psychological states, rooted in the benefits of predicting others' behavior;[2] (3) problem solving in a social context, for example, how we should punish offenders; and (4) social practice learning, by positive and negative reinforcement, by trial-and-error, by conditioning, and by analogy ([26]: 9).

At this point the story converges with proposals that try to explain the "mystery of biological altruism" by the evolutionary conformation of *homo reciprocans*, the being who is capable of giving, as long as he or she can somehow receive (Hauser and Levy, among others). Here Churchland will launch a hypothesis that is common to other authors: "A plausible hypothesis is that the desire to extend to all humans the respect and dignity once more or less limited to small groups probably originates here" ([25]: 103). Levy also considers that, as evolution reinforces the dispositions that favor adaptation, it could be thought that it has endowed us with a feeling of benevolence that in the early stages of brain formation was limited to those close to us, but, as the context changed, would have become universal benevolence, although he adds that this is speculative ([53, 56]: 302–303). However, in this way of proceeding there are at least three presuppositions that do not come to light.

First of all, that local benevolence is going to become universal is by no means an empirical fact. This claim is at the moment refuted by the fact that attempts to improve people with drugs and brain interventions have arisen, as we said, on the understanding that education is not succeeding.

[2] At this point, the role of mirror neurons seems to be key, although it is a matter that is still under discussion. See Rizzolatti and Sinigaglia [70] and Jacoboni [51].

Second, naturalists try to discover the empirical origin of morality and moral values already having a conception of morality, that of their own society, that of a post-conventional morality of principles. Accordingly, they only pick up the data that is consistent with that conception. They then understand that morality will be a matter of care, cooperation, mutual help. And if the facts show that benevolence is limited by the nature of a reciprocating human being, ready to give whenever he or she can expect some return, from what has been called "indirect reciprocity," they suggest (although speculatively) that the step to universal solidarity will be taken in the future.

This way of proceeding would be legitimate if it were announced without ambiguity that although no supernatural concept has been used, they do rely on concepts that have emerged culturally in a humanity that has also learned morally. Concepts such as "dignity" or "unconditional value," which demand an attitude of respect for those to whom they are attributed, are part of a world of predicates that are not natural, but cultural—that do not describe facts, but which are necessary for us to understand ourselves.

This is the case of some work in which Churchland addresses the concept of dignity and ensures that it is a constructed concept, like all concepts, but unlike other concepts does not refer to facts [25]. And she is right. But what is incoherent then is to affirm that dignity is more respected in some instances than in others and that we must learn from those more respected instances, there is, however, no clarification of what is meant by dignity, or by freedom and equality, or why "bioculturally" we have opted for these values as superior to others. These are not supernatural terms, but neither are they natural. They are moral terms and belong to the world of moral values, which we have been shaping bioculturally. It is necessary to concretize them in specific practices in order to understand their meaning in depth; we need to count on them for guidance.

Third, the release of oxytocin by prosocial behaviors may produce pleasure, but so do other behaviors, such as revenge or victory. As Evers (among others) points out, our innate identity, neurally based in a species-specific way, disposes us to develop universal evaluative tendencies, such as self-interest, dissociation, selective sympathy, empathy, and xenophobia. Our neural identity as people makes us social, but also individualistic and self-projective ([33, 34]: 2). Although we have a natural capacity for empathy (ability to apprehend the mental states of others), sympathy (ability to care for others), and mutual assistance, human beings can also be described as self-interested and xenophobic ([34]: 5). This is not strange considering the need for cohesion with those close to us, and protection from strangers, which powerfully influenced the shaping of the brain. But it is also aporophobic because it predisposes us to take into account those who can give us something back in exchange for what we give them, and to exclude the poor [21]. We experience pleasure in attending to our social feelings, but also to those who support our interest.

How can we account, on the basis of facts, for the emergence of a universalist ethic, such as that which permeates the Universal Declaration of Human Rights of 1948, which is based on the notion of human dignity? Could it be that moral learning is biocultural and, therefore, draws on diverse sources, be they philosophical, religious or political, to generate the proposal of what we want and should be?

5 An Open Future

For all these reasons, perhaps it would be appropriate to change the strategy in neuroethical research. Perhaps, instead of using empirical methods at the beginning, and later trying to resort to philosophical ones, learn something from what the critical Kant proposes: instead of waiting for empirical experience to teach us about morality, build our concepts and force nature to answer the questions that we have raised from them in order to try to understand empirical phenomena.

In fact, that is what most of the authors working in neuroethics have done when trying to discover the neural basis of morality: They already have a concept of "morality" when trying to identify which empirical phenomena line up with it.

But it is also necessary to mark a clear difference in relation to Kant: to design this concept we rely on the experience acquired throughout our biosocial and cultural history. There is no concept that human beings could have constructed had we not been able to count on it, but in order to construct this concept we do not limit ourselves to what we have learned from a strictly biological evolution, which does not exist. The interpretative framework to which we resort does not arise from empirical findings, but from that natural and social history along which we have been forging our concepts: not from an empirical reason, and neither from a pure reason, but from an *impure reason* [10, 14, 15]. As Habermas rightly says, throughout history we have not been learning only technically, but also morally. Although ontogenetically individuals must find their own path.

Therefore, neuroethics, as the neuroscience of ethics, should be a joint task of *ethicists and neuroscientists* in which psychologists, experimental economists, anthropologists and evolutionary biologists would also be involved, thus prolonging that history of interdisciplinarity, which is the history of human wisdom ([75]: 33). This, then, will be their tasks:

(1) To ask about the cerebral bases of moral conduct, clarifying what morality consists of and the wosetrld of categories that be used to understand it.
(2) To try to discover the *basis of moral obligation*. In order to attribute moral conduct to a being, this being must have a bodily basis, among which is the possession of a brain and a nervous system, articulated with the rest of the body. But "basis" is not the same as "foundation," and "necessary condition" is not the same as "sufficient condition." However, the question is whether it is possible to answer the question of the *reasons for moral obligation*, to answer the question "why should I?" in the face of certain moral demands, on the

basis of these foundations alone. The question about the basis of obligation is not the same as the question about the bases that a being must have in order to be capable of moral life [16, 17].

(3) To resort to ethical theories that allow us to *interpret the meaning of neuroscientific data for moral life*. Knowledge is not a set of fragmentary contributions, but requires a framework from which to interpret the data. Without "back stories," it is impossible to interpret scientific data ([74]: 378). There is an interaction between the hermeneutic framework and what can be qualified as "relevant findings."

(4) To try to design an ethical framework coherent with the most relevant data from neuroscience, to ask in what aspects these data force us to rethink important elements of ethical theory, and to find out to what extent ethical theory forces us to orient behavior in certain directions, marking the path to evolution. The human brain is, fortunately, plastic.

Evidently, these tasks are appropriate to a form of knowledge that is constituted as fundamental neuroethics, rather than as applied bioethics.

References

1. Amor JR (2015) Bioética y neurociencias. Institut Borja de Bioética-Universitat Ramon Llull, Barcelona
2. Arana J (2015) La conciencia inexplicada. Biblioteca Nueva, Madrid
3. Aranguren JL (1994) Ética. Obras Completas II. Trotta, Madrid, pp 159–502
4. Blackburn S (2008) Response to Marc Hauser's Princeton Tanner lectures. http://www.phil.cam.ac.uk/swb24/PAPERS/Hauser.pdf
5. Blanco C (2014) Historia de la neurociencia. Biblioteca Nueva, Madrid
6. Bonete E (2010) Neuroética práctica. Desclée de Brouwer, Bilbao
7. Buchanan A (2011) Beyond humanity? Oxford University Press, Oxford
8. Capó M, Nadal M, Ramos C, Fernández C, Cela Conde C (2006) Neuroética, derecho y neurociencia. Ludus Vitalis XIV(25):163–176
9. Codina MªJ (2015) Neuroeducación en virtudes cordiales. Octaedro, Barcelona
10. Conill J (2006) Ética hermenéutica. Tecnos, Madrid
11. Conill J (2010) De la ley natural al universalismo hermenéutico. Pensamiento 248:227–244
12. Conill J (2015) Naturaleza humana en perspectiva biohermenéutica. Pensamiento 269:1249–1260
13. Conill J (2016) La intimidad corporal y sus bases neurobiológicas. Pensamiento 72(273):789–807
14. Conill J (2020) Intimidad corporal y persona humana. Tecnos, Madrid
15. Conill J (2021) Nietzsche frente a Habermas. Genealogías de la razón. Tecnos, Madrid
16. Cortina A (2010) Neuroética: ¿las bases cerebrales de una ética universal con relevancia política? Isegoría 42:129–148
17. Cortina A (2011) Neuroética y neuropolítica. Tecnos, Madrid
18. Cortina (2011) Neuroética: ¿ética fundamental o ética aplicada? Diálogo filosófico 80:205–224
19. Cortina A (coord.) (2012) Guía Comares de Neurofilosofía práctica. Comares, Granada
20. Cortina A (2016) La conciencia moral desde una perspectiva neuroética. De Darwin a Kant. Pensamiento 72(273):771–788

21. Cortina A (2022) Aporophobia. Why we reject the poor instead of helping them. Princeton University Press, Princeton, NJ and Oxford
22. Choudhury S et al (2009) Critical neuroscience: linking neuroscience and society through critical practice. BioSocieties 4(1):61–77
23. Choudhury S, Slaby J (eds) (2011) Critical neuroscience: a handbook of the social and cultural contexts of neuroscience. Wiley-Blackwell
24. Churchland PS (1991) Our brains, ourselves: reflections on neuroethical questions. In: Roy DJ, Wynne BE (eds) Bioscience and society. Wiley & Sons, New York, pp 77–96
25. Churchland PS (2008) Human dignity from a neurophilosophical perspective. In: Human dignity and bioethics. President's Council on Bioethics, Washington, pp 99–121
26. Churchland PS (2011) Braintrust. Princeton University Press, Princeton
27. Cranford RE (1989) The neurologist as ethics consultant and as a member of the institutional ethics committee. The neuroethicist. Neurol Clin 7:697–713
28. Damasio A (1994) Descartes' error: emotion, reason and the human brain. Pan Macmillan
29. Damasio A (2000) The feeling of what happens: body and emotion in the making of consciousness. Harvest Books, USA
30. Champagne FP, Curley JP (2005) How social experiences influence the brain. Curr Opin Neurobiol 15:704–709
31. Dean R (2010) Does neuroscience undermine deontological theory? Neuroethics 3:43–60
32. Di Francesco M (2007) Neurofilosofia, naturalismo e statuto dei giudizi morali. Ethics & Politics 9(2):126–143
33. Evers K (2009) Neuroéhtique. Odile Jacob, Paris
34. Evers K (2015) Can we be epigenetically proactive. In: Metzinger T, Windt JM (eds) Open Mind, vol 13. Frankfurt a.M., pp 1–21
35. Gazzaniga MS (2006) El cerebro ético. Paidós, Barcelona
36. García-Marzá D (2012) Neuropolítica: una mirada crítica sobre el poder. In: Cortina A (coord.), op. cit., pp 77–96
37. García-Marzá D (2016) Neuroética aplicada: las consecuencias prácticas del neuropositivismo. Pensamiento 72(273):881–900
38. Giménez JM, Sánchez-Migallón S (2010) De la neurociencia a la Neuroética. EUNSA, Pamplona
39. Glannon W (ed) (2007) Defining right and wrong in brain science. Dana Press, New York
40. Gracia J (2016) ¿Incurre la teoría del proceso dual del juicio moral de Joshua Greene en falacia naturalista? Pensamiento 72(273):809–826
41. Greene JD (2012) Del 'es' neuronal al 'debe' moral: ¿cuáles son las implicaciones morales de la psicología moral neurocientífica? In: Cortina A (coord.), op. cit., pp 149–158
42. Greene JD (2014) Beyond point-and-shoot morality: why cognitive (neuro)science matters for ethics. Ethics 124:695–726
43. Habermas J (2006) Entre naturalismo y religión. Paidós, Barcelona
44. Habermas J (2009) Philosophische Texte. Bd. 5. Suhrkamp, Frankfurt
45. Haidt J (2012) El perro emocional y su cola racional: un enfoque intuicionista social del juicio moral. In: Cortina A (coord.) Guía Comares de Neurofilosofía práctica. Comares, Granada, pp 159–215
46. Haidt J (2012) The righteous mind: why good people are divided by politics and religion. Pantheon Books, New York
47. Haidt J et al (2000) Moral dumbfounding: when intuitions finds no reason. Unpublished manuscript, University of Virginia
48. Haidt J, Hersh MA (2001) Sexual morality: the cultures and emotions of conservatives and liberals. J Appl Soc Psychol 31:191–221
49. Hauser MD (2006) Moral minds. Harper Collins Publishers, New York
50. Illes J (ed) (2006) Neuroethics. Defining the issues in theory. Practice and policy. Oxford University Press, Oxford
51. Jacoboni M (2009) Las neuronas espejo. Katz, Barcelona

52. Insel T, Fernald RF (2004) How the brain processes social information: searching for the social brain. Ann Rev Neurosci 27:697–722
53. Kitcher P (2014) Is a naturalized ethics possible? In: de Waal F, Churchland PS, Pievani T, Parmigiani E (eds) Evolved morality. The biology and philosophy of human conscience. Brill, Leiden, Boston, pp 245–260
54. Kucewski M (1998) Cauistry. In: Chadwik R (ed) Encyclopedia of applied ethics, academic press. San Diego, London, pp 423–432
55. Lavazza A, de Caro M (2010) Not so fast. On some bold neuroscientific claims concerning human agency. Neuroethics 3:23–41
56. Levy N (2007) Neuroethics. Challenges for the 21st century. Cambridge University Press, Cambridge
57. Libet B et al (1982) Readiness potentials preceding unrestricted spontaneous pre-planned voluntary acts. Electroencephalograph and Clinical Neurophysiology, 54:322–325
58. Libet B (1999).Conscious vs neural time. Nature 352(6330):27–28
59. Libet B (1999) Do we have free will? J Conscious Stud 6:47–57
60. Lipina SJ (2014) Neuroethical considerations of child poverty. In: Salles A, Evers K (coords.) The social life of the brain. Fontamara, Mexico, pp 67–102
61. Marcos A (2010) Filosofía de la naturaleza humana. Eikasia 35:181–208
62. Marcos A (2015) Nuevas perspectivas en el debate sobre la naturaleza humana. Pensamiento 71(269):1239–1248
63. Marcus SJ (2002) Neuroethics: Mapping the field. Conference proceedings. The Dana Press, New York
64. Moll J et al (2005) The neural basis of human moral cognition. Nat Rev Neurosci 6(10):799–809
65. Mora F (2007) Neurocultura. Alianza, Madrid
66. Moreno J (2006) Mind wars. Brain research and national defense. DANA Foundation, New York
67. Morgado I (2012) Cómo percibimos el mundo. Ariel, Barcelona
68. Ortega C (2016) ¿Naturalizar la idea de justicia? Una respuesta crítica desde la teoría moral de Jürgen Habermas. Pensamiento 72(273):827–848
69. Richart A (2016) El origen evolutivo de la agencia moral y sus implicaciones para la ética. Pensamiento 72(273):849–864
70. Rizzolatti G, Sinigaglia C (2006) Las neuronas espejo. Paidós, Barcelona
71. Rorty R (2006) Born to be good. The New York Times, 27/08, pp 1–3
72. Roskies AL (2002) Neuroethics for the new millennium. Neuron 35:21–23
73. Schlag S (2013) Speculation and justification in policy-making on neuroenhancement. Recerca 13:11–28
74. Slaby J (2011) Perspectiven einer kritischen Philosophie der Neurowissenschaften. Deut Z Philos 59:375–390
75. Suhler C, Churchland PS (2011) The neurological basis of morality. In: Illes J, Sahakian BJ (eds) The Oxford handbook of neuroethics. Oxford University Press, Oxford, pp 33–58
76. Zubiri X (1986) Sobre el hombre. Alianza, Madrid

Ethics of Sport: Origins, Main Approaches, and Future Challenges

8

Francisco Javier López Frías

Abstract

This chapter comprehensively surveys the principal philosophical currents and traditions in sport ethics. It traces the historical evolution of the schools of thought that shaped this emergent philosophical sub-discipline, critically analyzing their strengths and inherent limitations. Particular emphasis is placed on the three main normative theories of sport: formalism, conventionalism, and interpretivism (or broad internalism). The chapter commences by situating the origins of these normative accounts of sport within early scholarly examinations of the relationships between play, games, and sport (dubbed the "tricky triad"). Subsequently, from this foundation, it chronicles the progressive development of these normative frameworks, beginning with formalism and explaining how this theory gradually gave way to conventionalism and, later, the emergence of broad internalism as the most widely embraced normative theory of sport.

Keywords

Sport ethics • Formalism • Conventionalism • Interpretivism • Play • Games

1 The Ethics of Sport: Is Sport a Serious Activity Worthy of Philosophical Reflection?

The publication of Huizinga [15] *Homo Ludens: A Study of the Play-Element in Culture* was one of the earliest catalysts for the scholarly study of the philosophical and cultural aspects of sport. Although Huizinga did not focus exclusively on

F. J. López Frías (✉)
Pennsylvania State University, University Park, PA, USA
e-mail: fjl13@psu.edu

A. Cortina and M. Correa Casanova (eds.), *Applied Ethics from Medicine to Humor*, Integrated Science 21, https://doi.org/10.1007/978-3-031-59009-2_8

sport but rather on the broader concept of "play" [32, 51], his irruption into the intellectual and academic landscapes is essential for understanding the emergence of the philosophy of sport in general and the ethics of sport in particular. Huizinga showed that "play and culture are actually interwoven with one another [because] pure play is one of the main bases of civilisation" ([15]: 5), encouraging numerous scholars to focus on the most prevalent and meaningful play activity today: sport. Huizinga was a great inspiration to many luminaries now considered precursors of the philosophy of sport, including Suits, Weiss, Fraleigh, Kretchmar, and Zeigler, among others.[1]

As Huizinga indicates, the omnipresence and predominance of sport in daily life are undeniable. Many people know much more about sport than other activities usually conceived as more important, including politics, economics, and education. To be sure, a vast majority of ordinary citizens can recite by heart their lifelong team's sporting achievements or line-ups but struggle to do the same when it comes to their basic rights as recognized in their countries' constitutions. Sport is a central activity in the lives of people from all nationalities, ages, and cultures. For example, a recent study on European citizens' sporting habits shows that 40% of those interviewed practice sport at least once a week.[2] According to the *2015 Sports Habits Survey,* in Spain, not only do 46% of the population engage in weekly sporting activities, but a significant 81% of citizens have attended a sporting event at least once a year. Additionally, 61% of individuals stay informed about sport through various media outlets, including newspapers and audiovisual platforms.

Besides the statistics, sport holds a unique symbolic power in society. Prompted by Geertz's classic studies on cockfighting as "deep play" ([13]: 339–378), historians, anthropologists, and philosophers have pondered the symbolic value of sport, that is, its potential to give meaning to people's lives both individually and collectively. In this way, sport is a "tool" [11] to, among other things, generate a common will, find a universal language through which to break down cultural and linguistic barriers, and exercise "soft power" ([14]: 407–418). For example, the United States has been configured as a country through its sports practices, especially baseball and football, forming what Dyreson [11] has called a "sporting republic." Similarly, the current Chinese government has made soccer a central axis not only of its educational system, making it a compulsory subject, but also

[1] *Homo Ludens* also influenced thinkers like Ortega y Gasset and Fink, who devoted a part of their works to reflecting on sport [23]. Although, curiously, Huizinga ([15]: 275, note 11) refers to Ortega y Gasset's text entitled "The Sporting Origin of the State" as one of his inspirations.

[2] European Commission, "Sport and physical activity," 2014. Available at http://ec.europa.eu/commfrontoffice/publicopinion/archives/ebs/ebs412enpdf. Accessed 12 May 2017.

of the country. Xi Jinping's government aims to use the prestige linked to victories on the soccer pitch with the country's economic success.[3] A social practice as impactful and socially relevant as sport cannot escape the rigorous scrutiny of philosophy and ethics.

2 The Ethics of Sport: The Most Applied Sub-discipline in the Philosophy of Sport

Among the branches of the young philosophy of sport, which became a discipline in its own right in the 1960s, the ethics of sport is undoubtedly the most widely cultivated and visible. Many publications in the two leading journals dedicated to the philosophy of sport, *Journal of the Philosophy of Sport* and *Sport, Ethics and Philosophy*, directly relate to or discuss ethical issues affecting sport. Similarly, the first journal in Spanish devoted to this subject, *Fair Play, Revista de Filosofía, Ética y Derecho del Deporte*, has a strong emphasis on ethics. Moreover, the first monograph of a Spanish philosophy journal on the philosophy of sport, published by *RECERCA: Revista de Pensament y Analisi*, explored the ethical challenges of sport in the twenty-first century [5]. To further support the prevalence of ethical examinations in the philosophy of sport, the most famous book series within the discipline, edited by Routledge, is *Ethics and Sport*. More recently, Oxford University Press promoted the philosophical reflection around sport by publishing two books of ethical content: *The Ethics of Sport: What Everyone Needs to Know* by Simon [49] and *The Ethics of Sport. Essential Readings*, edited by Caplan and Parent [4]. In fact, Simon's text is a reworking of one of the most classic texts within the discipline: *Fair Play*: *The Ethics of Sport* [45, 50].

According to McNamee [29, 30], the shift toward addressing applied ethical issues within the philosophy of sport, which is already inherently applied, resulted from the exhaustion of the definitional analytical debates that marked the discipline's early stages. This interpretation of the origins and history of the ethics of sport puts this scholarly field on a par with other applied ethics, which emerged from the hermeneutic-applied turn in Western philosophy during the first half of the twentieth century [7, 8, 21]. Thus, a leading cause of the inception of the ethics of sport was the reception of MacIntyre's (in particular, his virtue ethics framework in *After Virtue*) and Dworkin's hermeneutic philosophies [24]. Moreover, just as bioethics identifies the Nuremberg Trials, the emergence of resuscitation methods, and the first organ transplants as the foundational events that "knocked on the door of philosophical-ethical reflection," several foundational events during the mid-twentieth century "knocked on the door" of philosophy, human movement

[3] C. Jim and A. Jourdan, "Soccer schools and boardroom deals: China's play for World Cup glory | Reuters." Available at http://www.reuters.com/article/us-soccer-china-idUSKBN0TP0JF20 151206. Accessed 12 May 2017.

sciences, and sport education departments, spurring the philosophical examination of sport. Among these events are the Vietnam War, in particular, Muhammad Ali, John Carlos, and Tommie Smith's, among others, use of sport as a tool for protest and political activism, as well as the transfer of the political and social confrontation between the United States and the Soviet Union during the Cold War to the sporting arena.

Early philosophers of sport, influenced by Anglo-American analytical philosophical frameworks, concentrated on shedding light on the relationship among play, game, and sport, known as the tricky triad. However, conceptual clarity was necessary and developed as a response to practical concerns in the academic fields of human movement sciences and sport education [16]. Therefore, the philosophy of sport has always focused on practical issues [18, 21]. In addition to defining essential concepts such as "game," "competition," "victory," and "test," the founders of the philosophy of sport, inspired by classical thinkers such as Aristotle and Kant and phenomenologists such as Husserl, tackled fundamental ethical questions concerning the good life and how sport helps to achieve it.

Reflections about the place of sport in human life have always been at the heart of the philosophy of sport. The hermeneutic turn, which resulted from the reception of MacIntyre's and Dworkin's works, expanded the applied scope of the field, encouraging those who cultivated it to examine applied dimensions beyond the initial boundaries the discipline's founders set. One such dimension was the ethical foundations of sporting practices and their institutions to provide athletes and sport administrators with guidance on addressing questions concerning the elucidation, protection, and promotion of the values of sport. Fraleigh's *Right Actions in Sport: Ethics for Contestants* [12], Arnold's *Sport, Ethics and Education* [2], and McNamee and Parry's *Ethics and Sport* [28] mark the beginning of the ethics of sport as a sub-discipline of the philosophy of sport.

3 Main Trends in the Ethics of Sport

3.1 From Descriptive Statements to Normative Evaluations, a Good Start?

The main normative (i.e., ethical) approaches to sport are formalism, conventionalism, and interpretivism (or broad internalism). Their emergence within the philosophy of sport must be understood in conjunction with early sport philosophers' definitional efforts to capture the nature of sport. Initially conceived to elucidate and delineate the defining elements of sport, these theories were later deployed to provide ethical guidance [17].

3.2 Formalism

Formalism, as an ethical theory of sport, resulted from applying [53] definition of games and sport to normative issues. For him, sports are games of physical skill that enjoy a certain level of stability and a wide following.[4] By including the stability and following aspects, this definition separates sport from mere physical activity or physical exercise. For example, running a marathon organized by a sporting institution, built around universally accepted rules, and followed by a broad audience is considered sport. In contrast, running around the neighborhood for health reasons is not a sport despite its many similarities with running a marathon. However, the key to understanding the nature of sport and, above all, its ethical potential does not lie in the last two terms of the definition ("institutional stability" and "wide following"), but in the first notion: "game."

In *The Grasshopper: Games, Life, and Utopia,* Suits responds to Wittgensten's assertion that the concept of game cannot be defined by proposing the following definition:

> [A game is the] attempt to achieve a specific state of affairs [prelusory goal], using only means permitted by the rules, where the rules prohibit the use of more efficient in favour of less efficient means [constitutive rules], and where the rules are accepted just because they make possible such activity [lusory attitude]. ([52]: 41)

Rules are the cornerstone of games. They create the artificial world or "*intermezzo*"—using Huizinga's term—in which games occur. Consider the example of golf. The state of affairs to be achieved is to put a ball into a hole. Attempting to bring about this situation would hardly be a game if people used the most efficient means available, for instance, if they put the ball in the hole by hand. Games require the overcoming of challenging obstacles. To create such conditions, rules restrict the use of more efficient means in favor of less efficient ones, requiring the use of specific skills, such as hitting the ball with a golf club, to achieve the goal of the sport.[5] Rules, therefore, provide games with their basic formal structure, defining their goal and how to achieve it. This is why Suits calls them

---

[4] In "Tricky Triad: Games, Play, and Sport," a work subsequent to "The Elements of Sport" in which Suits initially defines sport as a game of physical skill, sport is defined as "competitive events involving a variety of physical (usually in combination with other) human skills, where the superior participant is judged to have exhibited those skills in a superior way" ([51]: 2). This later definition eliminates the term "game," allowing Suits to distinguish between two types of sport practices: refereed games and judged performances. This modification to the definition raised controversy, resulting in Meier's rejection of the claim that not all sports are games. Meier's view of sport, which aligns with Suits' original definition, is now the most widely accepted among sport philosophers. This is not to say that this is the only definition of sport. For example, Møller or Eichberg have offered broader definitions that include games and other types of physical activities.

[5] The Royal & Ancient Golf Club, "The rules of golf and golf equipment. The game. General rule I." Available at http://www.randa.org/Rules-of-Golf/MainRules/1-The-Game/SubRules/1-General. Accessed May 13, 2017.

"constitutive rules." Without them, there would be no games. Given the fundamental character of constitutive rules, playing a game is impossible if participants violate them. Put differently, anyone who wants to play a game *must* abide by the rules. That is to say, players must accept the rules *for the game to be possible.* Those who break the rules cease to be players.

The formalist notion of constitutive rule traces back to Kant and Searle [44]. Most philosophers of sport have drawn on the latter to conceptualize the formal structure of games further. For Searle, the world contains two types of facts: (a) institutional, which are constructed, and (b) brute, which are found in the natural, physical world. Searle [43] calls realities constructed through the creation and acceptance of rules "institutions." Within them, the brute facts of reality are given a distinct and exclusive meaning specific to the institution. For example, outside a golf match, putting a ball into a hole is a meaningless action. The action only becomes meaningful when pursued within the context of golf, where it is considered the primary goal of the sport.

For Searle, brute facts impose certain physical limitations that humans must accept. For example, "given humans' bodily configuration, they cannot hit the golf ball at a speed greater than X." These limitations are very different from institutional ones. The latter arise from obligations generated through establishing and accepting rules that determine what individuals can and must do within a particular practice. Searle identifies two types of rules: "constitutive" and "regulative." The former stipulate the defining characteristics of activities, creating the activity as such. Thus, a constitutive rule in golf would be: "Hit a ball with a club until it is driven into a hole in the fewest number of attempts." On the other hand, regulative rules do not define the activity but prescribe specific directions on how to perform it: "Moving the ball with the hand to make it easier to hit is penalized with two attempts."

Since the creation and acceptance of the rules of golf generate the meaningful context of the sport, the normative implication of formalism is that, to operate within the institution, participants *must* accept and play by the rules. These obligations are at the core of what Suits first and Morgan later call the "logical incompatibility thesis":

> one cannot win, let alone compete, in a game if one resorts to cheating. This is so … because in an important sense the rules of the game are inseparable from its goal … if one cannot really win a game unless one plays it, and if one cannot really play a game unless one obeys its rules, then it follows that winning and cheating are logically incompatible. ([35]: 1)

Per this thesis, playing and cheating are inherently incompatible, as breaking the rules entails a fundamental disruption of the formal structure of a game ([39]: 358–359). However, in sport, violating some rules, especially for strategic reasons, is commonplace and regarded as legitimate. For instance, in basketball, players frequently foul opponents to prevent them from making an easy basket or, when the fouled player is a poor free-throw taker, to stop the clock, send the player to the free-throw line, and regain possession.

Formalists have tried to reconcile such rule-breaking practices with the incompatibility thesis. Suits, for instance, explains that different rules have diverse functions. While some make up the formal core of the activity, others have a secondary role. The former establish the conditions that must be followed to play the game and include, by way of extension, a subcategory of rules that can be broken because they are linked to a specified penalty. Hence, Suits accepts the strategic violation of some rules. Although he does not label such rules, other philosophers of sport, such as D'Agostino [9], Meier [32], and Reddiford [40], have called them "regulative rules." Serving as extensions of constitutive rules, regulative rules protect the stability of a game by identifying ways for players to continue playing or restore the gameplay when a rule violation occurs. Take the example of soccer; handball rules are regulative because they specify how to continue or restore the play after a participant breaks a core constitutive rule of the game: "The ball must be driven with any part of the body other than the arms."

Some rules apply to in-game actions; others regulate the conditions that must precede a game or sport competition, such as who can enter the event and under what conditions (e.g., having benefited from a specific technology). Thus, Meier proposes a third category of rules: auxiliary rules. These are of great ethical relevance, as they typically lead to the exclusion and discrimination of participants. Some of the most widely-discussed ethical debates in sport revolve around auxiliary rules: the Paralympic athlete Oscar Pistorius' request to compete in the London 2012 Olympic Games, the controversy surrounding sex testing practices to which some sporting institutions subjected women athletes, the ban on transgender women from participating in women's competitions, and the penalization of athletes for violating World Anti-Doping Agency's rules.

The logical incompatibility thesis is the result of drawing normative judgments from ontological analyses of sport. As a normative thesis, it identifies actions aligning with the defining elements of games as morally acceptable and actions undermining such elements as morally objectionable. However, deriving normative content from descriptive judgments is controversial.[6] Some philosophers of sport deny that the logical incompatibility thesis has normative content [31]. For instance, for Moore, "Suits does claim that if you want to play a game then you *must* follow its rules, but this 'must' is logical, not moral. If you don't follow the rules you aren't playing the game—but it doesn't follow that you are doing anything morally wrong" ([34]: 104). In a similar vein, López Frías argues that the normative content of Suits' proposal is not located in the logical incompatibility

[6] Deriving normative statements from descriptive analyses, however, results in what philosophers refer to as the naturalistic fallacy, which raises doubts about the validity of formalism as a normative theory [19].

thesis but in his utopian world comprising individuals who devote their lives to playing games, which has a normative-regulatory function similar to that of the Kingdom of Ends in Kant's ethics [25].[7]

The formalist normative approach to sport has been widely criticized. One area of criticism that receives attention is the function of rules in games. Philosophers of sport, such as Kretchmar (2005), argue that addressing ethical issues by merely appealing to rules is problematic. Identifying these theoretical problems led to the development of sport philosophical theories (i.e., conventionalism and interpretivism) that complement rule-based analyses with elements beyond the rules, including social conventions and normative principles like excellence. Among the ethical challenges faced by formalism, two stand out. The first is that rules cannot cover all possible cases; the rulebook remains incomplete. The following cases illustrate this limitation [19]:

(a) Actions unforeseen by game designers. For example, the United States National Basketball Association (NBA) had to modify the dimensions of the basketball court, specifically the "zone" (i.e., the restricted area under the basket), to counteract the dominance of Wilton Norman "Wilt" Chamberlain. The tremendous influence this player's height and strength had on the game had not been contemplated when its rules were enacted.
(b) Actions whose evaluation varies depending on the context or game situation. Continuing with the example of professional basketball, the application of traveling rules (i.e., "a player may only take two steps without dribbling") varies across different contexts. In the NBA, referees frequently disregard these rules to increase the probability of spectacular dunks and athletic maneuvers. In contrast, traveling rules are applied more severely in basketball competitions regulated by the International Basketball Federation, such as the Basketball World Cup.
(c) Actions that cannot be regulated. In some cases, the decision as to whether there has been a rule violation depends on the intentionality of the rule violator. However, assessing the voluntariness or fortuitousness of an action is difficult. Consider the handball rule in soccer. Historically, regulatory bodies have failed to provide an uncontroversial formulation of the rule, mainly because they have sought to sidestep the evaluation of the players' intentionality. Nevertheless, the different formulations of the rule and their application consistently provoke contention, as it appears inevitable to adjudicate instances of handball violations without considering, in one way or another, the violator's intentionality.
(d) Actions of the game whose regulation would be inefficient or counterproductive. No soccer rule requires teams to stop playing or throw the ball out of

[7] Expanding on these issues goes beyond the scope of this book chapter.

bounds when a peer falls on the ground due to a physical problem. Players simply kick the ball out of bands, allowing the medical team to enter the field and assist the injured player.

The second problem formalism encounters is that rules cannot justify, change, or interpret themselves. Additional critical elements are needed to demonstrate why rules are necessary, how they must be modified, and when they must be applied or ignored. For example, as explained above, basketball rules dictating the specific dimensions of the zone were modified because extant rules failed to promote the kind of play game designers intended, specifically, that scoring resulted from team play involving pass combinations and ball movement.

3.3 Conventionalism

Chronologically speaking, conventionalism is the second ethical approach to sport. It seeks to overcome the limitations of formalism (Hager, 2014: 344–346). In fact, many have dubbed conventionalism as "anti-formalism" due to its origin as a critique and reformulation of formalism. In this sense, D'Agostino, creator of conventionalism, states that this philosophical approach to sport must preserve the valuable elements of formalism: "my point is not that formalism cannot be salvaged as an account of games. It is, rather, that if formalism is to be salvaged, it must be supplemented, one way or another, by an account of the ethos of games" ([9]: 16).

D'Agostino proposes to complement formalism through what he calls "ethos of the game," that is, "an unofficial *system* of conventions which determines how the official rules of the game will be applied in various concrete circumstances" ([9]: 14). Appealing to such a system of conventions, according to D'Agostino, allows formalist philosophers of sport to overcome two fundamental theoretical limitations: the rigidity of the (constitutive) rules and the Platonic view of games resulting from reducing them to their formal structure ([9]: 9). Regarding the former, D'Agostino's conventionalist proposal criticizes the logical incompatibility thesis. Since this thesis stipulates that participants who break the rules cease to play a game, formalism seems unable to account for rule-breaking actions typically considered acceptable and, most importantly, part of the game.[8] The clearest example of such actions is strategic fouling, i.e., voluntary violations of the rules to gain an advantage [34]. Fouling an opponent about to score an easy basket or in a one-to-one situation with the goalkeeper are commonly used and widely accepted intentional fouls. They have become a part of the game, as basketball and soccer players have agreed to make them so.

Per D'Agostino, players "conspire to ignore the rules" on certain occasions, such as to create opportunities for using strategic thinking to gain an advantage.

[8] This critique ignores formalists' distinction between constitutive and regulative rules.

According to the founder of conventionalism, resorting to these conventions allows for a more flexible understanding of game rules and their application. From a rigid formalist standpoint, a rule violation would render a game impossible, for the artificial space in which it occurs is created by the players' rule adherence. In contrast, conventionalists posit that players usually violate the rules without being disqualified or the game breaking down. Indeed, players typically differentiate "fair play" from "unsporting play." Despite disregarding the rules, unsporting play is still an instance of play. The player who displays unsporting behavior is still considered a player. For instance, someone who makes a strategic foul in a soccer match is accused of unsporting play but not of not playing the game.

The rigidity of rules ties with the second major criticism of conventionalism to formalism: its inherent Platonism. According to D'Agostino, formalism's excessive rigidity in rule application generates an ideal vision of games that cannot be realized in any specific instance of the game, much less in the highly competitive spheres of professional sport. The pressure to compete, the physical demandingness, and the play intensity make rule violations inevitable. This criticism does not only refer to cases such as intentional rule violation but also to situations involving unintentional rule violations due to the very nature and dynamics of the game (e.g., a soccer player risks a difficult tackle to steal the ball but takes down their opponent). Even the "protectors" of the rules, the referees, choose to "conspire against the rules" on certain occasions. For example, umpires in a baseball game with poor pitchers who throw too many balls often expand the batting area to facilitate more *strikes* and speed up the game.

Despite overcoming some of the identified limitations of formalism, conventionalism has its critics, specifically Simon [46], Dixon [10], and Butcher and Schneider [3]. The conventional nature of the ethos of sport is their primary target. For them, the conventions that comprise the ethos of a game lack critical power; the mere existence of socially agreed actions involving the modification or suspension of some rules should not confer such actions normative value. For example, that most professional athletes accept the simulation of fouls as a valid game strategy does not morally justify this practice. Moreover, conventions produced by specific sporting communities often stem from non-sporting interests or concerns inherent to the community, rendering a game vulnerable to non-game-related forces such as economic or political interests. Take the traveling fouls example presented earlier. NBA referees are more permissive with these fouls to increase the league's spectacle value. Spectators want to see remarkable plays. To promote these, referees often suspend traveling rules, even if this suspension arguably negatively affects the game by encouraging individualistic play and reducing team passing and long-range shooting.

Ciomaga, a leading conventionalist, argues that D'Agostino's work aims to describe rather than provide a normative theory. Ciomaga highlights that the creator of conventionalism explicitly states that his proposed theory of sport aims to determine, through an empirical investigation of the specific practices that conventions sanction, what conventions direct people's actions and referees' decisions [6]. However, Morgan, the foremost defender of conventionalism, accepts the normative drift of conventionalism. His conventionalist proposal, "deep conventionalism" or "ethical conventionalism," starts from the premise that both defenders and detractors of D'Agostino's original conventionalism hold an extremely superficial understanding of the nature of conventions [36]. By drawing on Marmor's [27], Lewis' [20], and, in the latest formulation of his conventionalist framework, Hegel's views of conventions, Morgan delves deeper into the concept to provide conventionalism with a more solid foundation, allowing it, in his opinion, to solve the criticisms directed against it [38].

According to Morgan, conventions are ubiquitous and plural. Everything humans do depends on conventions of some kind. Morgan distinguishes between two types of conventions: coordinate and deep [36]. Following Lewis, he conceives the former as agreements reached collectively to respond to regular cooperation problems. An example of this type of convention is traffic regulations. Coordinating different individuals driving vehicles in the same space is a prevalent social problem to which individuals must find a collective solution. Thus, they agree to drive on the same side of the road and to stop when a traffic light is red. Which color determines who should stop or what side of the road people drive on are contingent and arbitrary choices; however, the fact that traffic circulation becomes difficult, if not impossible, without such conventions is neither contingent nor arbitrary. The circulation problem is a reality with which individuals must deal collectively.

Following Marmor, Morgan states that most conventions are not of a coordinated type but rather play a more decisive role in people's lives and social practices. These types of conventions are "deep" or "ethical." According to Morgan, deep conventions ground the activities in which people engage. The problem of driving a vehicle in shared space exists regardless of people's attitudes. However, many human activities cannot exist separate from people's conventions. Rather, such conventions help create and shape the activities. One such activity is games or, more narrowly, sport. That is to say, in games and sport, deep conventions, like rules, have a constitutive function.

Sport activities are meaningless without deep conventions. These are not mere pragmatic agreements to coordinate action, but, together with rules, they constitute and shape the nature of sport activities, particularly the values, excellence, and skills inherent to them. Hence, for Morgan, deep conventions have a normative role. Certainly, the emergence of sport responds to certain immutable human traits, such as bodily types and basic needs, which, like the problems coordinate conventions address, exist independently from human conventions. However, deep conventions shape how individuals perceive such immutable factors and feature them in their social practices, leading to the development of those practices in different forms depending on individuals' conventions [36].

Concerning the plurality of deep conventions and, in turn, sporting communities, Morgan refers to the two most prominent understandings of sport in Western history: amateur and professional [37]. Both views of sport depend on different deep conventions. Amateurism considers sport a pastime or leisurely activity (i.e., one engaged for its own sake) centered on celebrating people's natural talents and hard work. The amateur athletes' sporting skills and, especially, achievements must spring from their effort and inborn abilities alone. Moreover, amateur athletes must embody a chivalrous and noble attitude focused on cooperating to help one another excel. Thus, they experience sport through deep conventions that identify sports as autotelic activities aimed at promoting excellence, which contrasts sharply with professional sport's emphasis on achieving victory (at almost any cost).

Professionalism turns sport into a serious activity far removed from the autotelic playfulness attributed to it in the amateur ethos. Professional athletes are most concerned with the rewards that come with winning, such as social recognition, power, and money. In this sense, instrumental thinking and performance enhancement are crucial in the professional world. Moreover, the focus on instrumental pursuits and victory hinders athletes' ability to value sport as an end in itself. Taken to an extreme, the professional ethos is captured in the adage erroneously attributed to the American football coach Vince Lombardi: "Winning isn't everything, it is the only thing" ("winning isn't everything, but wanting to win is") ([50]: 37).

The amateur and professional understandings of sport build upon profoundly different sets of conventions, raising cooperation problems when athletes from each sporting community interact. Butcher and Schneider ([3]: 11) illustrate this kind of problems through the case involving the implementation of the penalty area in soccer. Amateur soccer players opposed this change because it would radically change the sport. In their view, it would not only complicate the task of refereeing matches but also provide players with greater incentives to seek to deceive referees to be awarded a penalty kick. Deceiving referees, those in the amateur band argued, should not be rewarded in soccer. In contrast, professional soccer players embrace the penalty area, as it gives them greater opportunities to gain a competitive edge. Indeed, in today's soccer, some attacking players specialize in drawing fouls in the penalty area, and deception skills have become "intrinsic" means to outsmart opponents.

Deep conventions coordinate participants' actions and, most importantly, constitute sporting practices by shaping how athletes perceive them. Thus, Morgan believes that deep conventionalism escapes the criticisms of D'Agostino's conventionalism, especially the charge that it lacks critical capacity. Following Rorty and Kuhn, Morgan ([37]: 49) argues that a critical examination of social practices and the deep conventions grounding them is not only possible but also a vital part of such practices. Given the contextual character of deep conventions, Morgan explains that the critical examination of practices can only take place internally, that is, from "within the walls" [22] of each given practice, because rational dialogue is only feasible among those sharing the same view of the social practice, that is, those relying on the same deep conventions to engage in discussions about the nature of the activity. Otherwise, rational dialogue and critical examination become impossible. This hardly means that amateur and professional athletes share nothing in common; it implies that they cannot rationally agree on the fundamental issues affecting their different understandings of sport ([22]: 46). According to Morgan, rational dialogue across different practice communities is impossible. The professional soccer player, for instance, may question the suitability of deception skills. However, for Morgan, this will not happen through rational dialogue with an amateur player but as a result of encountering practical or theoretical problems in professional soccer. Yet, transcontextual exchanges may be beneficial, as they spur rational reflection; that is, they may encourage individuals within a practice community to reflect upon and criticize the deep conventions.

In sum, Morgan's deep conventionalist proposal, contra anti-conventionalist attacks, holds that deep conventions have critical power. However, he rejects that such critical edge be grounded in using universal or transcontextual rational criteria aligning with something called the nature or essence of sport. Morgan calls the attempt to devise such rational critical principles "the view from nowhere," using Nagel's famous concept. Pursuing certain defining principles specific to sport is precisely the goal of the third ethical approach to sport, interpretivism or broad internalism.

3.4 Interpretivism or Broad Internalism

Broad internalism is "the view that in addition to the constitutive rules of sports, there are other resources connected closely—perhaps conceptually—to sport that are neither social conventions nor moral principles imported from outside" ([46]: 7). This philosophical approach to sport is termed "broad" because it appeals to a set of elements inherent or internal to sport practices beyond the rules and conventions. Put differently, sport interpretivists acknowledge that rules and conventions are defining elements of sport. However, they add a third essential trait: normative principles that belong intrinsically—and, in many cases, exclusively—to sport. According to Simon, these "principles must be presupposed in order for us to arrive at the most comprehensive, coherent, and morally acceptable account of [sport] itself, the best interpretation of [sport]" ([50]: 34). Simon's interpretivist approach is the most prominent normative theory of sport. Among others, Simon bases his proposal on Dworkin's interpretivist philosophy of law, which, to grasp the nature of the law better, goes beyond written rules to identify principles without which the law is meaningless. In this way, the law is interpreted by appealing to principles internal to it rather than to moral or political principles imported from outside ([50]: 36).

According to Simon, three types of broad internalism can be differentiated: (a) contractualism, (b) respect for the game, and (c) mutualism. These three proposals share the key idea that one cannot fully understand sport and provide normative guidance without identifying normative elements beyond the rules and conventions. However, they differ in their understanding of the origins and nature of the normative element to be identified. The contractualist approach to sport holds that sport practices are grounded in an implicit social contract among the participants. By entering this contract, players agree to engage in a particular practice and abide by both the rules and the conventions that make it possible. That is to say, before agreeing on how to play the game (by adhering to rules and conventions), they must demonstrate their intention to engage in a common activity. Proponents of this approach point to the opening ceremonies before sporting events as the symbolic moment when the pact is sealed. In Olympic sport, for example, countries parade around the stadium carrying their respective flags in front of one another and the spectators. Similarly, in international soccer, at the start of a match, teams face each other (or the audience) and respectfully listen to each other's national anthem; then, the players and coaching staff shake hands. These ceremonies display the chronological order of the contract: First, the players present themselves in the same place and express their willingness to engage in the game. Subsequently, they greet each other, recognizing themselves as opponents. Finally, the teams (or captains) greet the referees as a sign of respect for the rules. In ancient Greece, the contractual agreement was made explicit verbally. Competitors gathered in front of the statue of Zeus and recited an oath similar to the Hippocratic oath that medical doctors must swear before they start practicing medicine ([33]: 120).

The respect for the game approach to sport starts from a place similar to the contractualist one. As formulated by Butcher and Schneider, this normative theory of sport defends that participants follow specific rules and conventions because they have agreed to respect the "interests" of the game. Thus, in reaching this agreement, players adopt the goals or interests of the game as their own. This proposal draws heavily from Kant. Acknowledging the importance of respecting the interests of the game requires conceiving the game as an intrinsically valuable entity that, in the style of the Kantian rational being, demands respect [3]. To further identify the interests of games, Butcher and Schneider rely on two other philosophical proposals: (a) Kretchmar's theory of contests as events wherein participants compare their performances and (b) MacIntyre's concept of social practice. In this way, a game is seen as an event in which two individuals, or teams, test each other within a level playing field to discover who is better at exercising specific skills and, in doing so, realize certain goods internal to the practice ([3]: 17).

Finally, mutualism is based on a view of sport as "*a mutually acceptable quest for excellence through challenge*" ([50]: 46, original emphasis). Although Weiss [55] and Fraleigh [12] already described sports in these terms, Simon and those who defend his mutualist proposal give the view of sport as a mutual quest for excellence a greater dimension, making it the interpretive key to understanding better sport and the fundamental normative criterion to evaluate sporting practices.

According to Simon, the emphasis on competition has been taken to the extreme today's sport. This has turned sport competitions into zero-sum games or non-cooperative games. In this type of game, only the winner gets something out of their engagement in the activity. Those who are defeated get nothing. In sports, winners are typically celebrated and praised, while losers are despised and criticized. Thus, it is hardly surprising that athletes concentrate their efforts on seeking competitive advantage to beat their opponents. After all, if they fail to beat their opponents, their efforts go unrewarded. In Simon's mutualistic approach to sport, competition is an essential defining element. Without competition, participants would not try their best. Therefore, sport competitions would fail to help identify who is best at exercising specific skills. Simon [48] criticizes the importance given to competition in today's extremely competitive sports. According to him, sports need not be zero-sum games. Instead, they must be understood as cooperative quests for excellence. This view of sport highlights sport's pedagogical and cooperative characteristics, overshadowed by those views of sport that conceptualize the activity solely in competitive and oppositional terms.

Mutualism regards sport as a pedagogical and humanistic enterprise, similar to that found in Olympism and its founder, de Coubertin [42]. From a mutualist standpoint, sport is not defined by the pursuit of victory but is primarily linked to the realization of internal goods related to human flourishing, that is, to the realization of human capacities [1]. Sport, therefore, is seen in an Aristotelian fashion: it is a perfectionist-based, autotelic activity. Thus, when confronting ethical issues in sport, Russell, another proponent of broad internalism, states that "*rules should be interpreted in such a manner that the excellences embodied in achieving the lusory goal of the game are not undermined but are maintained and fostered*" ([41]: 35, original emphasis).

According to Morgan, the major drawback of interpretivism is adopting a "view from nowhere" of sport, one centered on eliciting excellence, which hardly fits with the plurality of sporting traditions and communities, each with different understandings of sport. Referencing another concept mentioned above, interpretivism provides a Platonic view of sport, an ideal that does not align with the reality of sport. However, this critique of interpretivism fails, as it ignores the hermeneutic roots of this sport philosophical proposal. The conception of sport as a mutual quest for excellence need not rely on a fixed, eternal, and idealistic essence of sport. Simon argues that mutualism provides the best and most ethical interpretation of sport. Other interpretations can be achieved, but the mutualist one brings out the best elements in sport, turning it into a pedagogical and humanistic practice that helps players to realize themselves as human beings. In fact, Simon draws on Apel and Habermas' discourse ethics to counter those who criticize him of essentialism, affirming that in his mutualist proposal:

> Discourse ... might originate in a particular social and historical context, but to the extent that it has normative force, it is based on prin-ciples and can lead to conclusions that purport to be something more than a mere historical consensus. To the extent that the "something more" can in principle be understood in terms of intellectual cogency and merit, it goes beyond reflective ethnocentrism toward realism. ([47]: 138)

For Simon, mutualism is a view of sport that can be arrived at from diverse interpretations of sport through rational dialogue. This does not mean that all sporting communities would eventually harbor the same conception of sport, but it does mean that the view of sport as a mutual quest for excellence would be at the core of the multiple understandings of sport—when they reach a rational agreement on the nature of sport.

Returning to the accusation of Platonism but interpreting internalism through hermeneutical lenses, López Frías and Gimeno Monfort criticize mutualism for providing an unclear and intellectualized vision of excellence [26]. These critics argue that mutualists must clarify the kinds of excellence acquired through sport, specifying whether they have a moral component or are solely bodily excellences. Similarly, López Frías critiques Simon's mutualism for severing elements intrinsic to sport from those external to it ([23]: 362–364; [54]). In many cases, ethical problems affecting sport cannot be solved exclusively by drawing on normative elements connected to sport as a struggle for excellence. However, engaging with broader normative concepts, such as coercion, autonomy, and dignity, is necessary in other cases. See the recent debate around the inclusion of transgender athletes in competitive sport (i.e., in the category that best aligns with their gender identity), wherein the key lies in whether a categorization system can balance the promotion of competitive fairness and broader justice-related goals concerning the inclusion of historically marginalized social groups.

4 Conclusion: A Young but Vigorous Discipline

Although the ethics of sport is a young discipline, this chapter has shown that the various philosophical approaches to sport that comprise it draw from strong and long-standing theories, including, but not limited to, virtue ethics, phenomenology, pragmatism, and hermeneutics. This gives the discipline a solid theoretical foundation and tools to tackle theoretical and practical problems affecting sport. In doing so, the ethics of sport offers critical tools to promote better sporting practices and, from a more utopian standpoint, envision and lead better lives by engaging in those practices. The latter has always been a central ethical concern for philosophers of sport and sport ethicists. It is no coincidence that one of the founding works in these disciplines, *The Grasshopper,* contains the terms "games," "life," and "utopia" in its subtitle. As Suits puts it, a fundamental ethical challenge is to ponder whether the ants in Aesop's fable are wrong in declining the grasshopper's invitation to build their lives around play-centered activities, such as games and sport, instead of work.

References

1. Anderson D (2001) Recovering humanity: movement, sport, and nature. J Philos Sport 28(2):140–150
2. Arnold PJ (1998) Sport. Ethics and education. Cassell, London
3. Butcher R, Schneider A (1998) Fair play as respect for the game. J Philos Sport 25(1):1–22
4. Caplan AL, Parent B (2017) The ethics of sport: essential readings. Oxford University Press, London
5. Chiva Bartoll Ó, López Frías FJ (2016) Current perspectives on the philosophy and pedagogy of sport." *Recerca*, n° 18, pp. 7-12.
6. Ciomaga B (2013) Rules and obligations. J Philos Sport 40(1):19–40
7. Cortina A (2008) Ética aplicada y democracia radical. Tecnos, Madrid
8. Cortina A, García-Marzá D (2003) Razón pública y éticas aplicadas: los caminos de la razón práctica en una sociedad pluralista. Tecnos, Madrid
9. D'Agostino F (1981) The ethos of games. J Philos Sport 8(1):7–18

10. Dixon N (2003) Canadian figure skaters, French judges, and realism in sport. J Philos Sport 30(2):103–116
11. Dyreson M (1998) Making the American team: sport, culture, and the Olympic experience. University of Illinois Press, Urbana
12. Fraleigh WP (1984) Right actions in sport: ethics for contestants. Human Kinetics Publishers, Champaign, IL
13. Geertz C (1983) La interpretación de las culturas. Gedisa, Barcelona
14. González L (2008) Soft y 'blando': contagio léxico y empobrecimiento semántico. In: González L, Hernúñez P (eds) Traducción, contacto y contagio: Actas del III Congreso 'El español, lengua de traducción'. Esletra, Brussels. https://cvc.cervantes.es/lengua/esletra/pdf/03/038_gonzalez.pdf
15. Huizinga J (1950) Homo ludens. A study of the play-element in culture. Alianza Editorial, Madrid
16. Kretchmar RS (1997) Philosophy of sport. In: Messengale J, Swanson R (eds) The history of exercise and sport science. Human Kinetics, Champaign, IL, pp 181–202
17. Kretchmar RS (2007) Dualisms, dichotomies and dead ends: limitations of analytic thinking about sport. Sport Ethics Philos 1(3):266–280
18. Kretchmar RS (2014) The philosophy of sport and analytic philosophy. In: Torres CR (ed) Bloomsbury companion to the philosophy of sport. Bloomsbury, London, pp 41–52
19. Kretchmar RS (2015) Formalism and sport. In: McNamee M, Morgan WJ (eds) Routledge handbook of the philosophy of sport. Routledge, New York
20. Lewis D (2013) Convention. A philosophical study. Wiley, Hoboken
21. López Frías FJ (2011) Philosophy of sport: origins and development. Dilemata 5:1–19
22. López Frías FJ (2014) The philosophy of sport today. Paradigms and currents. (Qua.Pe.G) Università degli Studi di Roma Foro It, Rome
23. López Frías FJ (2014) Internalism. In: Torres CR (ed) The Bloomsbury companion to the philosophy of sport. Bloomsbury Publishing, London, pp 362–364
24. López Frías FJ (2015) The impact of Alasdair MacIntyre's *After Virtue* on the philosophy of sport. The equivocations of the internalist paradigm. Cuadernos Salmantinos de Filosofía 42:179–202
25. López Frías FJ (2017) A Kantian view of Suits' Utopia: 'a kingdom of autotelically-motivated game players. J Philos Sport 44(1):138–151
26. López Frías FJ, Gimeno Monfort X (2015) A hermeneutical analysis of the internalist approach in the philosophy of sport. Phys Cult Sport Stud Res 67(1):5–12
27. Marmor A (2009) Social conventions: from language to law. Princeton University Press, Princeton, N.J
28. McNamee M, Parry SJ (1998) Ethics and sport. Routledge, London
29. McNamee M (2007) Sport, ethics and philosophy; context, history, prospects. Sport Ethics Philos 1(1):1–6
30. McNamee M (2008) Sports, virtues and vices: morality plays. Routledge, London
31. Meier KV (1985) Restless sport. J Philos Sport 12(1):64–77
32. Meier KV (1988) Triad trickery: playing with sport and games. J Philos Sport 15(1):11–30
33. Miller SG (2012) Arete: Greek sports from ancient sources. University of California Press, Berkeley
34. Moore E (2017) Formalism and strategic fouls. J Philos Sport 44(1):95–107
35. Morgan WJ (1987) The logical incompatibility thesis and rules: a reconsideration of formalism as an account of games. J Philos Sport 14(1):1–20
36. Morgan WJ (2012) Broad internalism, deep conventions, moral entrepreneurs, and sport. J Philos Sport 39(1):65–100
37. Morgan WJ (2015) Conventionalism and sport. In: Morgan WJ, McNamee MJ (eds) Routledge handbook of the philosophy of sport. Routledge, New York, pp 35–53
38. Morgan WJ (2020) Sport and moral conflict. A conventionalist theory. Temple University Press, Philadelphia

39. Pérez Triviño JL (2014) Formalism. In: Torres CR (ed) The Bloomsbury companion to the philosophy of sport. Bloomsbury Publishing, London, pp 358–359
40. Reddiford G (1985) Constitutions, institutions, and games. J Philos Sport 12(1):41–51
41. Russell JS (1999) Are rules all an umpire has to work with? J Philos Sport 26(1):27–49
42. Schantz OJ (2016) Coubertin's humanism facing post-humanism—implications for the future of the Olympic Games. Sport Soc 19(6):840–856
43. Searle JR (1997) The construction of social reality. Free Press, London
44. Searle JR (2011) Speech acts: an essay in the philosophy of language. Cambridge University Press, Cambridge
45. Simon RL (1991) Fair play: sports, values, and society. Westview Press, Boulder, CO, p 1991
46. Simon RL (2000) Internalism and internal values in sport. J Philos Sport 27(1):1–16
47. Simon RL (2004) From ethnocentrism to realism: can discourse ethics bridge the gap? J Philos Sport 31(2):122–141
48. Simon RL (2014) Theories of sport. In: Torres CR (ed) The Bloomsbury companion to the philosophy of sport. Bloomsbury Publishing, London, pp 83–98
49. Simon RL (2016) The ethics of sport: what everyone needs to know. Oxford University Press, London
50. Simon RL, Torres CR, Hager PF (2015) Fair play: the ethics of sport. Westview, Boulder, CO
51. Suits B (1988) Tricky triad: games, play, and sport. J Philos Sport 15(1):1–9
52. Suits B (2005) The grasshopper: games, life and utopia. Broadview Press, Peterborough, ON
53. Suits B (2007) The elements of sport. In: Morgan WJ (ed) Ethics in sport. Human Kinetics, Champaign, IL, pp 9–19
54. Thompson D, López Frías FJ (2020) Flourishing as the standard for evaluating the social practice of competitive sport. DILEMATA. Int J Appl Ethics 33:185–197
55. Weiss P (1971) Sport: a philosophic inquiry. Southern Illinois University Press, Carbondale

Ethics of the City

9

Mauricio Correa Casanova

Abstract

This chapter reviews the current situation of the urbanization process and discusses its ethical pathologies. It then analyzes three proposals in which the city is currently considered from an ethical perspective: the right to the city, the just city, and the ethical city. Given the need for a more adequate systematization of these approaches, the hermeneutic-critical alternative of an ethics applied to the city is proposed.

Keywords

City • Urban question • Just city • Right to the city • Ethical city • Ethic of the city

We live in an urbanized planet. Currently, more than 50% of humanity lives in urban areas, and projections indicate that by the year 2050 the concentration will reach 70% of the planet's inhabitants. Undoubtedly, the city has become the human habitat par excellence. This phenomenon has been analyzed from different disciplinary fields, but it still represents a pending task for applied ethics. In this chapter I propose a brief exposition on what I consider a new disciplinary interface for ethics. Namely, what I refer to as the "ethics of the city" [22].[1]

---

[1] Another approach to this line of research is called *Urban Ethics*. See the works of Mostafavi [72], Chan [18] and Ege and Moser [37]. In a recent book, Sennett [87] speaks of an "Ethic for the city."

M. Correa Casanova (✉)
Pontificia Universidad Católica de Chile, Santiago, Chile
e-mail: mcorrea9@uc.cl

A. Cortina and M. Correa Casanova (eds.), *Applied Ethics from Medicine to Humor*, Integrated Science 21, https://doi.org/10.1007/978-3-031-59009-2_9

The general objective of an ethics of the city lies in reconstructing the ethical discourse on the city ([10]: 55). In this endeavor it is not simply a matter of returning to the *polis/civitas*, a subject widely dealt with by philosophers, but requires broadening the gaze towards the *urbs*.[2] In this sense, the ethics of the city is constructed on the basis of the interdisciplinary dialogue established between moral philosophy and contributions from sociology, geography, architecture, urban planning, and not forgetting, of course, citizens, local and national governments, as well as businesses. Although the fruits of this collaborative work are yet to be seen, I foresee that the ethics of the city will very soon begin to occupy a place of privilege in the wide world of contemporary applied ethics alongside environmental ethics, development ethics, business ethics, bioethics, or the ethics of the professions [21]. In this context, the present chapter represents a first step or a sketch pending further systematization [22].

This chapter is structured as follows: first, I present a brief overview of the current state of the urbanization process and the need for ethics; second, I briefly dwell on a multidisciplinary look at the city; third, I analyze three proposals in which the city is currently posed in ethical perspective, namely, the "right to the city," the "just city," and the "ethical city"; fourth, I offer a critical hermeneutic approach to the development of an ethics applied to the city; fifth, I conclude with a projection on the urban question as an ethical revolution.

1 An Overview of the Urbanization Process

As a generalization, the city is the geographical space delimited and adapted to the conditions of collective human life. It contains material, institutional, intellectual, and spiritual provisions that, at least in principle, are intended to facilitate the unfolding of human life: houses, buildings, parks, commerce, networks of basic services (electricity, drinking water, internet, telephone, and television connections), hospitals, educational centers, churches, consumer goods and services, etc. The city brings together all the devices required for the unfolding of life, in such a way that it can be considered as the ecological sphere in the human sense. As such, it is a collective human work or construction, although this does not necessarily imply a participatory and democratic construction. It involves architects, urban planners, engineers, politicians, businessmen, and ideally also citizens. Despite this, the city is far from representing a homogeneous reality and, therefore, it cannot be reduced to a univocal understanding. At present, there is no standardized international criterion for determining the boundaries of a city, and there are often multiple different definitions of boundaries for a given city (e.g.

[2] In general, here I understand *polis/civitas* as referring to the sociohistorical being and its civic-political status; while *urbs* represents the physically constituted city in which the first one develops spatially, that is, the socio-spatial being and its urban status [17].

administrative center, urban agglomeration, or metropolitan area). However, there is no such thing *as a* city, but only many multifaceted constructions that give it concrete life. That is why we can affirm that it is embodied in many ways, as its history demonstrates [47, 73].

The passage from the rural to the urban, which is mediated by the industrialization process characteristic of the nineteenth century, is what defines the irruption of the city as a geospatial or urban phenomenon [64]. This process has been driven today to the point where there is no longer any doubt that we inhabit an urbanized planet [4, 89, 92]. According to the data provided by the *World Cities Report 2016, Urbanization and Development: Emerging Futures*, of the United Nations Human Settlements Programme [94], the world has significantly increased the concentration of its population in urban areas: from an annual average of 57 million between 1990 and 2000, to 77 million between 2010 and 2015. In 1990, 43% (2.3 billion) of the world's population lived in urban areas, while in 2015 this increased to 54% (4.3 billion). Projections suggest that by 2050 the figure will rise to between 66 and 70%. This increase in the urban population has not been homogeneous throughout the world. Asia, for example, has the largest number of people living in urban areas.

Some effects of this accelerated urbanization process are notorious. According to the same WCR 2016, cities are responsible for generating up to 70% of greenhouse gas emissions (especially through the consumption of fossil fuels used for energy supply and transport). In a boomerang movement, cities not only contribute to climate change, but are also vulnerable to its worst effects. Moreover, the growth of slums or urban informal settlements, especially in developing countries, is prominent. The Global Report on Human Settlements 2003, entitled *The Challenge of Slums*, describes the housing conditions of the majority of the urban poor, who survive in precarious conditions through informal settlements and informal income-generating strategies [93]. In this regard, Wacquant [95] deals with the "urban outcasts" and marginality in the city at the beginning of the millennium, and Mike Davis, in his book *Planet of Slums*, provides abundant and overwhelming data on the precarious conditions of the "hyperdegraded urban areas" [33]. This is compounded by the difficulty of accessing adequate housing, the deficiency of basic urban services, road congestion, and other factors such as growing social exclusion and inequality, insecurity and violence, as well as the increase in international migration.

Against this general backdrop, discourses on the inclusive, safe, resilient, and sustainable city (SDG 11) appear as mere rhetoric, as the reality of cities reveals an urban development that at this point is humanly and ecologically unsustainable, vulnerable, exclusionary, and unequal. A great human concentration out of control and with a disorganized, bureaucratic (without urban citizenship), and mostly privatized (commodified) urban planning has given rise to what we can call the *tragedy of the cities* (a new version of the tragedy of the commons denounced by Garret Hardin in 1968). In practice, all this comes to question the ethical conquests of modernity ([1]: 5), such as the recognition of the dignity of every human being and the human rights that stem from this concept, as well as the firm ethical conviction that we should respect and care of the natural environment, acquired from the second half of the twentieth century. Undoubtedly, these and other ethical conquests have not been integrated into the very process of urban transformation and planning. However, precisely because of this observation, in recent decades sociologists, geographers, architects, and urban planners have become increasingly interested in incorporating an ethical dimension into the discussion on cities. In this sense, as the outstanding geographer and urban planner Jordi Borja stated in an interview that today "[t]he perhaps most important challenge is to reconstruct the ethical or moral discourse on the future city, the desired city" ([10]: 55). An ethics of the city, therefore, appears as necessary and urgent in order to counteract the unjust consequences of the process of urbanization at the global level.

2 Multidisciplinary Views on the City

Since the time of the Greek philosophers, the *polis* (city-state) has been the subject of philosophical reflection [13]. The constitutive reality of humans as "social beings" (*zõon politikon, animal socialis*) is the central element around which the city is philosophically understood. Along this line, Plato wrote his brilliant treatise on the *Republic* and the *Laws*, as well as Aristotle *The Politics* [90]. Later, in the patristic period, Saint Augustine distinguished between the "earthly city" and the "eternal city" in his work *De civitate Dei* [50]. In the Renaissance, the city was the place par excellence of utopian thought with Thomas More (*Utopia,* 1516), Tomaso Campanella (*The City of the Sun,* 1623), and Francis Bacon (*New Atlantis,* 1623) [34]. In a certain way, city and utopia would also be the subject of Karl Marx and Friedrich Engels, although this time in the perspective of the industrial city and the revolution of the proletariat already exhausted from the eschatological wait for the city of God [15]. Heidegger [57] would also leave his mark on the city in his hermeneutic reflections on the *Bauen, Wohnen, Denken*, a post-war conference that was first exhibited in Darmstadt in 1951, and which is considered by some of its followers as an authentic paradigm of the philosophy of contemporary architecture [85].

This tour (very partial, by the way) represents the outline of what we can consider a certain *philosophy of the city.*[3] Its main feature consists in the fact that it is an ontological and at the same time totalizing approach; that is, the city appears as an abstract representation of metaphysical and anthropological type—the sociotemporal being that is projected in the political and the religious. One must consider, however, the important nuance of the reflections offered by Engels

[3] For a collection of philosophical texts on the city, see Ansay and Schoonbrodt [2], Meagher [71], Pereira [82, 83], among others.

in *The Situation of the Working Class in England*, of 1845 and the *Contribution to the Problem of Housing*, of 1873 [38, 39], and perhaps to a lesser extent, Heidegger in the context of the German post-war reconstruction.

In this development, nothing is said about the city as place, space, or territory that is configured as a collective construction, symmetrical and equitable, in which important processes of production, planning and distribution intervene. Without minimizing its extraordinary contributions in the field of social and political philosophy, it can be affirmed that for philosophers and philosophy the city (*urbs*) has not been an objective condition of the socio-spatial being, an external datum with its own material specifications ([61]: 46). We can include in this new line of inquiry the recent, valuable contributions of Young [98], Nancy [76] and Hénaff [59].

In the field of sociology, contributions on the city (*urbs*) have a longer trajectory.[4] Its beginnings go back to Max Weber and his *Die Stadt. Eine Sociologische Untersuchung,* first published in 1921 in the *Archiv für Sozialwissenschaft und Sozialpolitik* ([19]: 363). In this text Weber defines the city in a spatial sense as a "locality," in an economic sense as a "market place," and in an administrative political sense as an "urban municipality" ([96]; also [97]: 938–1046). In Weber's words:

> One can try to define the "city" in many different ways. All definitions will have, however, a common element: the city does not consist of one or many separately settled buildings, but, on the contrary, constitutes a concentrated (or, at least, relatively concentrated) habitat, a "locality" .... We will speak of the "city," in the economic sense, when the population residing in a locality satisfies an economically substantial part of its needs in the local market, thanks mainly to the products that this population, and that of the immediate surroundings, have manufactured or have procured to sell them in the market. Every city, in the sense that we have conferred here to this word, is a "market place," that is to say, every city has as economic center of the settlement a market .... It is convenient in any case to remember that the economic concept of "city," up to here exposed, must be clearly distinguished from the administrative political concept of "city," since only from this last perspective a delimited urban territory belongs to the city. ([96]: 3, 5 and 14–15)

Robert E. Park (1864–1944), the distinguished founder of the Chicago School, developed the so-called human ecology of urban sociology studies. Of great importance is his article entitled "The City: suggestions for the investigation of human behavior in the city environment," originally published in 1915 in the *American Journal of Sociology*, later revised and published in *The City* of 1925. For Park:

> The city .... is something more than a congeries of individual men and of social conveniences – streets, buildings, electric lights, tramways, and telephones, etc.; something more, also, than a mere constellation of institutions and administrative devices – courts, hospitals, schools, police, and civil functionaries of various sorts. The city is, rather, a state of mind, a

[4] For a history of "the sociologists of the city," see the interesting work of Bettin [8], an anthology in Bassols et al. [7] and Lin and Mele [66].

> body of customs and traditions, and of the organized attitudes and sentiments that inhere in these customs and are transmitted with this tradition. The city is not, in other words, merely a physical mechanism and an artificial construction. It is involved in the vital processes of the people who compose it; it is a product of nature, and particularly of human nature. ([81]: 1)

In the second half of the twentieth century, specifically, from the urban crisis of the 60 s, it would be Jane Jacobs (1916–2006) in her work *The Death and Life of Great American Cities*, 1961, who was the first to launch an attack "on the principles and aims that have shaped modern, orthodox city planning and rebuilding" ([60]: 3). In her critique, Jacobs refers mainly to the economic reasons of the supposed urban rehabilitation, which she considers a "deception" if its most visible and palpable effects are appreciated:

> Low-income projects that become worse centers of delinquency, vandalism and general social hopelessness than the slums they were supposed to replace. Middle-income housing projects which are truly marvels of dullness and regimentation, sealed against any buoyancy or vitality of city life. Luxury housing projects that mitigate their inanity, or try to, with a vapid vulgarity. Cultural centers that are unable to support a good bookstore. Civic centers that are avoided by everyone but bums, who have fewer choices of loitering place than others. Commercial centers that are lackluster imitations of standardized suburban chain-store shopping. Promenades that go from no place to nowhere and have no promenaders. Expressways that eviscerate great cities. ([60]: 4)

In Jacobs' opinion, all this "is not the rebuilding of cities," but rather the "sacking."

In this context of crisis also appears the neo-Marxist variant of urban studies, which creates a new paradigm of study on the city. In this tradition, the French sociologist and philosopher, Henri Lefebvre (1901–1991), occupies a place of distinction. Lefebvre wrote *Le droit à la ville*, a work originally published in 1968, which marks a turn in thinking about the city, and which has recently been vindicated by the efforts of those who try to understand the pathologies of urban space built on the dynamics of neoliberal capitalist logic.[5] For our author, the city is presented "as a *projection of society on the ground*, that is to say, not only on the sensible space but on the specific plane perceived and conceived by thought, which determines the city and the urban" ([61]: 75).

Among Lefebvre's many contributions [61–64], we will highlight here at least two of his main ideas. One is his distinction between "use-value" and "exchange-value." In his opinion, the thriving reality of the city "is, in itself, *work*, and this

[5] Some of these recent and abundant publications on the right to the city include the following: Goonewardena et al. [51], Nehls et al. [77], Sugranyes and Mathivet [91], De Mattos and Link [35], Costes [30], Gasic et al. [49] and Mathivet [70]. See also *Urban* magazine, no. 2 (2011), dedicated to "Specters of Lefebvre," with the collaboration of Peter Marcuse, Jean-Pierre Garnier, Thierry Paquot, among several others.

characteristic contrasts with the irreversible orientation to money, to commerce, to exchange, to *products*. In effect, the work is use value and the product, exchange value" ([61]: 18. 45). Hence his fundamental thesis:

> The city and urban reality are revelatory of use value. The exchange value, the generalization of merchandise through industrialization, tend to destroy, subordinating to it, the city and the urban reality, refuges of use value, germs of a virtual predominance, and of a revaluation of use.
>
> The action of these specific conflicts is exercised in the urban system that we intend to analyze: between use value and exchange value; between the mobilization of wealth (in money, in paper) and the unproductive investment in the city; between the accumulation of capital and its waste in parties; between the extension of the dominated territory and the demands of a severe organization of this territory that outlines the dominating city ([61]: 20).

In short, Lefebvre's thesis is that the city and urban space are becoming increasingly commodified and commercialized, according to the logic of their exchange value.

Another, and Lefebvre's most valuable contribution and the idea that has had the greatest success, is that of the right to the city [61, 65]. Although we will return to it later, for Lefebvre the right to the city "is announced as a call, as a demand" and is incarnated as "the *right to urban life*, transformed, renewed" ([61]: 138). Although Lefebvre makes a broad use of the expression [32], we can synthesize his thought in the following statement:

> The right to the city manifests itself as a superior form of rights: the right to freedom, to individualization in socialization, to habitat, and to inhabit. The right to the *work* (to the participating activity) and the right to *appropriation* (very different from the right to property) are imbricated in the right to the city. ([61]: 159)

In this regard, Deneulin comments:

> Lefebvre proposed to recover the use value of cities as social, cultural and political meeting places. For this it was essential [to have] the participation of all residents of urban areas … in the decisions that affected their lives, and in the creation of new urban spaces valued by the use that their inhabitants would give them. ([36]: 256)

In this way, the neo-Marxist vision of the city and urban space is realized through a dialectic that relates work, use value, and appropriation.

Along with Lefebvre, the urban sociologist Castells [16] also stands out in this line of the neo-Marxist tradition, especially his work *La Question Urbaine*, of 1972. Also of note is the geographer Harvey [54] and his work *Social Justice and the City*, published in 1973. With both authors, penetrating analyses on the internal dynamics of the capitalist logic stand out and present a series of urban pathologies. The capitalist configuration of urban space rests on a neoliberal individualist ethic that promotes the accumulation of wealth and consumption, which at the same time generates a perverse dynamic of overcrowding, poverty, dispossession, marginalization, inequality, and gentrification. Taken together, these authors' theories can be considered the "Marxification" of contemporary urban knowledge, as Soja ([88]: 100) calls it, which has been dominant in the last decades.

It is precisely in this last and recent neo-Marxist theoretical elaboration where the moral concern about city studies appears. In opposition to the positivist tendency in social sciences, the moral dimension manifests itself as a denunciation of the social injustice that is nested in the urban projects of the neoliberal capitalist city. As Susan Fainstein argues:

> Even while not explicitly developing a concept of justice [86], they *attacked injustice within cities* and sought to demonstrate how spatial manipulation underlay the disadvantages suffered by residents of poor neighborhoods. Operating on both a theoretical and empirical level, urban scholars on the left contested conventional analyses and *injected a moral dimension into their work.* ([42]: 1. Italics mine)

Despite this important innovation, Fainstein states that none of these authors explicitly developed *a normative concept of justice.* But neither, we might add, did they rely on a broader normative ethical framework, either to address the ethical pathologies of the city, or to offer ethical guidelines to contribute to a more just urban development.

3 Contemporary Ethical Approaches to the City

There are currently at least three ethical approaches to the city, namely, the right to the city, the just city, and the ethical city. In the following, I will briefly outline each of them in terms of their main representatives.

3.1 The Right to the City

The Right to the City, as we saw, is part of the neo-Marxist tradition whose main author is Henri Lefebvre. His book has undoubtedly contributed to an upward evolution on the appropriation of the city, and to a new understanding of urbanism today. In fact, today this ideal is highly influential; it has been strongly claimed by many authors from different fields (but especially by sociologists and geographers) who deal with urban issues, as well as by urban policy actors and a variety of civil organizations at local, regional, and international levels [31]. It should be noted, however, that the claim for the right to the city has not always been strictly situated within the coordinates of Lefebvrian thought. In this sense, according to David Harvey, today "the right to the city is an empty signifier. Everything depends on who gets to fill it with meaning" ([55]: xv).

According to the ethical and political model of human rights, the right to the city is for Harvey a "collective right," as is the case with the rights of workers, women, LGBTQ + people, and other minorities ([55]: 3). Specifically:

> The right to the city is … far more than a right of individual or group access to the resources that the city embodies: *it is a right to change and reinvent the city more after our heart's desire.* It is, moreover, a collective rather than an individual right, since *reinventing the city inevitably depends upon the exercise of a collective power over the processes of urbanization.* ([55]: 4. Italics mine)

The right to the city, therefore, is expressed collectively as a claim to "some kind of shaping *power over the processes of urbanization*, over the ways in which our cities are made and remade, and to do so in a fundamental and radical way" ([55]: 5. Italics mine). In this sense, for our author the right to the city is one of the most neglected human rights. This is mainly because the process of urbanization has sprouted from the geographical and social concentration of a surplus in production that as such is integrated into the totality of the capitalist system of perpetual accumulation. Capitalist development and the process of urbanization are intimately connected. Following Marx's explanation, capitalism rests "upon the perpetual search for surplus value (profit). But to produce surplus value capitalists have to produce a surplus product. This means that capitalism is perpetually producing the surplus product that urbanization requires. The reverse relation also holds. Capitalism needs urbanization to absorb the surplus products it perpetually produces" ([55]: 5). The fundamental problem is that this process takes place entirely on the basis of class division and the power of use of a few, the justification for which is found in the neoliberal individualist ethic that has dominated in recent decades. From this perspective, any attempt to absorb the surplus through urban transformation, what has been called "creative destruction," creates a vicious circle, since it possesses a class dimension in which "usually the poor, the underprivileged, and those marginalized from political power that suffer first and foremost from this process" ([55]: 16). In short, and according to this interpretation, cities as new urban geographies have inevitably been built on a horrific dynamic, which is contrary to the collective power over the process of urbanization.

But if this is the situation, then what should be demanded by the vindication of the right to the city? In Harvey's view, if the neoliberal project has been oriented by the privatization of control over the surplus, the answer is quite simple:

> … greater democratic control over the production and use of the surplus. Since the urban process is a major channel of use, then the right to the city is constituted by establishing democratic control over the deployment of the surpluses through urbanization. ([55]: 22–23)

In this case, if the surplus falls under state control, this is only of any use "if the state itself is both reformed and brought back under popular democratic control" ([55]: 23) and not in the hands of the private interests of a small political and economic elite "who are in a position to shape the city more and more after their own particular needs and heart's desire" ([55]: 24). The right to the city as a collective right, therefore, is simultaneously a working slogan and a political ideal by which the process of urbanization and its connection to the production and use of surplus is reinvested. In this tenor, according to Harvey, perhaps Lefebvre was right more than half a century ago to insist that "the revolution in our times has to be urban-or nothing" ([55]: 25, see [48, 100]).

Another influential author who has insisted on the right to the city is Borja [9, 11]. In his opinion, in our time, gender, environmental, communicational, and collective identity rights have emerged and along with them the so-called right to the city. In his own words:

> The right to the city integrates rights related to the physical environment, such as housing, public space, transportation, the environment, etc., which condition individual and collective rights of a social or political nature, that is, the effectiveness of the status of the citizen. But also integrated into the right to the city are political and social rights that in turn condition insertion in the city, such as political and legal equality, the personal and collective identity of minorities, the citizen's salary or basic income, continuous training, etc. *The right to the city, then, is a set of formal and material rights that configure citizenship.* ([11]: 145. Italics mine)

From this perspective, there is a close relationship between the right to the city and citizens' rights.

> We choose to conceptualize these citizen rights as citizens and not "human" because we consider that they are part of the status of citizenship. It is to recognize the person as a subject of rights and duties that make him or her free in the territory in which he or she has chosen to live and equal to all those who live in this territory. ([11]: 145–46)

Therefore, city and citizenship are intertwined in the same struggle, which consists of making legitimate citizens' rights effective in our urbanized societies.

Thus, Borja proposes a list, although not exhaustive, of "urban rights" as a contribution to the political culture in the sphere of the city and local government. The list is as follows:

- Right to housing and place
- Right to public space and monumentality
- Right to beauty
- Right to collective identity
- Right to mobility and accessibility
- Right to centrality
- The right to the conversion of the marginal or illegal city into a city of citizenship.
- Right to metropolitan or multi-municipal government
- Right to political innovation
- Right of access to and use of information and communication technologies
- Right to the city as a refuge

- The right to protection by the local government from higher political institutions, and service-providing organizations and companies.
- Right to local justice and security
- Right to illegality
- Right to employment and citizen's salary
- Right to environmental quality
- Right to difference, privacy, and choice of personal relationships
- The right of all residents of a city to have the same legal political status as citizens.
- The right for direct representatives of citizens, both institutional and social, to participate in and have access to international conferences and organizations.
- Citizens' rights to equal mobility and access to cross-cutting information
- The right of local and regional governments, and organizations to form networks and associations ([11]: 146–152).

Although a list of rights may seem ineffective and inoperative, in Borja's opinion, each right fulfils a legitimizing function that seeks to influence the orientation of public policies. In its vindicatory task, a key issue will be to assume such rights not in isolation, but in a relationship of interdependence ([12]: 50 ff.).

Along these lines, the right to the city has been expressed in the adoption of several international charters, including: the *World Charter for the Right to the City*, proposed by popular movements, non-governmental organizations, and professional networks of civil society, based on an initiative that was born at the Earth Summit held in Rio de Janeiro in 1992 and continued in 2001 in Porto Alegre (Brazil) at the First World Social Forum;[6] the *European Charter for the Safeguarding of Human Rights in the City*, signed in Saint-Denis, France, in 2000;[7] the *World Charter-Agenda for Human Rights in the City*, a municipal charter of human rights of global scope by the UCLG Commission on Social Inclusion, Participatory Democracy and Human Rights, formally adopted in 2011, within the framework of its World Council in Florence.[8] There are two differences between the Charter-Agenda and the European Charter: their geographical scope (one is

---

[6] The World Charter is available [online]:

http://www.onuhabitat.org/index.php?option=com_docman&task=doc_details&gid=50&Itemid=3.

For their analysis and weighting see Nehls et al. [77] and Sugranyes and Mathivet [91].

[7] On the *website* of the world organization of cities, United Cities and Local Government (UCLG), it is reported that this European Charter "is the result of the preparatory work initiated in Barcelona in 1998 in the framework of the Conference 'Cities for Human Rights', which was organized in commemoration of the 50th Anniversary of the Universal Declaration of Human Rights." From 2008, the promotion of the European Charter was entrusted to UCLG, through the Commission on Social Inclusion, Participatory Democracy and Human Rights. The European Charter and the list of signatory cities are available [online]: https://www.uclg-cisdp.org/es/el-derecho-la-ciudad/carta-europea.

[8] The Charter-Agenda is available [online]: https://www.uclg-cisdp.org/es/el-derecho-la-ciudad/carta-mundial.

global, while the other is European); and the action plan for their implementation at the local level, which is only present in the former. There are also other local charters, e.g., the *Montreal Charter of Rights and Responsibilities* (Canada, 2006), and the *Mexico City Charter for the Right to the City* (Mexico, 2010). Finally, the right to the city is also enshrined in the national legislation of Brazil and Ecuador, the two countries that led efforts to include the concept in the UN-Habitat III process (Quito), albeit without much success.

Without going into a detailed analysis of each of these charters, it is sufficient here to refer to the *World Charter for the Right to the City*, which establishes that it is a new human right, according to which the right to the city is a human right:

> *All persons have the right to the city without discrimination as* to gender, age, health conditions, income, nationality, ethnicity, migratory status, political, religious or sexual orientation, as well as to preserve cultural memory and identity in accordance with the principles and norms set forth in this Charter. (I. 1. Italics mine.)

Then, in Art. I. 2, the World Charter defines the right to the city, which includes:

> [T]he equitable use of cities within the *principles of sustainability, democracy, equity and social justice.*
>
> It is a *collective right of city dwellers*, especially vulnerable and disadvantaged groups, which gives them legitimacy of action and organization, based on their uses and customs, with the aim of achieving the full exercise of the right to self-determination, and an adequate standard of living.
>
> *The right to the city is interdependent with all internationally recognized human rights, integrally conceived, and therefore includes all civil, political, economic, social, cultural and environmental rights that are already regulated in international human rights treaties* ….
>
> [T]he Right to the City also includes the right to development, to a healthy environment, to the enjoyment and preservation of natural resources, to participation in urban planning and management, and to historical and cultural heritage.

In short, the right to the city is based on human rights—not as one more of those rights, but as the right to enforce in the city the rights that already formally exist. It seeks to rescue the city from its mercantile use, as it is dominated by the interests of capital and economic growth, in order to conquer it for its citizens.

3.2 The Just City

Currently, one of the most interesting proposals is the so-called *just city* understood as the ultimate ethical objective of urban planning. A good sample of the discussions around this proposal is the volume *Searching for the Just City* (edited by Marcuse et al. [68]), which brings together a series of theoretical and applied analyses that were initially presented at a conference on urban justice at Columbia University [69]. The lead essay in this volume is by Fainstein [43], a specialist in urban planning, who is recognized as the most important promoter of this idea, especially for her book entitled *The Just City*, published by Cornell University Press in 2010.

Fainstein's analysis is limited only to what seems feasible within the current context of capitalist urbanization in rich, formally democratic Western countries ([40]: 5)—cities such as New York, London, or Amsterdam. Her analysis, therefore, does not extend to all cities in an abstract and universal way (i.e., it does not apply to poor, hyper-degraded cities). Within this context, Fainstein observes that the justifications that dominate capitalist urban planning projects are given in terms of investment, profitability, competitiveness, and economic growth, so that consequences concerning social impact and justice are rarely mentioned within the urban context. Her proposal, therefore, seeks to provide guidance on what to do if justice is the first evaluation criterion used in urban policy-making. In her own words, her aim "is to develop an urban theory of justice, and to use it to evaluate existing and potential institutions and programs" ([40]: 5), all within the framework of a critical view of capitalist urban policy that seeks to reduce both government intervention and that of citizen movements, while instead privileging market processes and economic growth.

In Fainstein's view, leftist urban scholars, especially beginning with Henri Lefebvre and all those who were influenced by him, rejected positivism and the guise of scientific rigor in order to inject a moral dimension to their work, both theoretically and empirically, by attacking injustice within cities. However, in much of the available urban literature this critique was undertaken without explicitly developing a normative concept of justice ([41]: 11, [42]: 1–2). Although David Harvey uses the word "justice" or "social justice" in the development of his analyses, most notably in 1973's *Social Justice and the City*, he does so in an ambivalent way, as he rejects what he sees as "idealized prerequisites" ([41]: 14). In this context, although our author expresses her agreement with Harvey in the sense that the content of the term *justice* assumes different meanings depending on the social, geographical, and historical context (2010: 23), she assumes the challenge of *making explicit* a normative definition of justice as a foundation for urban planning.

Unlike the neo-Marxist thinkers, Fainstein proposes her own point of view making use of the contributions coming from the contemporary moral and political philosophy, with Habermas [53], Young [98, 99], and Nussbaum [78, 79], among others. In particular, our author uses the proposal of the recognized American philosopher Rawls [84] and his influential theory of justice as fairness. In her view, "Although Rawls does not address justice in the city, his argument for a just society defined by greater equity stimulated much later work within political philosophy that in turn filtered into urban social science" ([42]: 6, [41]: 14). In particular, Fainstein focuses her attention on the important difference principle formulated by Rawls:

> It is well known that Rawls argued that the distribution of goods in a society should be governed by the "difference principle," whereby policies should only improve the situation of those in better circumstances when "doing so would be to the advantage of those less fortunate". ... Rawls proposes that in the "original position," behind a "veil of ignorance" that prevents individuals from knowing what their ultimate position in society will be, they will opt for a more egalitarian distribution of goods. According to this logic, egalitarianism is a rational choice of individuals. ([41]: 14)

From this perspective:

> If Rawls's conception of justice is applied to the city, fair distribution of benefits and mitigating disadvantage should be the aims of public policy. Rawls's use of the phrase "prevent excessive concentrations of property and wealth" implies a realistic utopianism - the expectation is not of eliminating material inequality but rather of lessening it. Thus, the criterion for evaluating policy measures, according to Rawlsian logic, is to insure that they most benefit the least well off. ([44]: 4)

Thus, if a given urban planning policy serves the objective of assisting the most disadvantaged, then that policy is to be considered efficient even if it does not contribute to maximizing the aggregate cost/benefit *ratio* derived from the utilitarian perspective ([41]: 15). If the role of the planner and his identification with this view of distributive justice is taken into account, then his proposals have to promote policies that favor the least favored or low-income groups. Only in this way, according to Fainstein [40], can it be considered fair urban planning.

In this context, and in response to the criticisms launched against Rawls' theory, Fainstein argues in *The Just City* that diversity, democracy, and equity are three distinctive criteria of urban justice that help in the evaluation of the just city. Along the lines of Iris Young and Nancy [45, 46], diversity requires recognition and respect for differences. Thus, among urban planners, diversity is defined by both the mix of land uses and the social heterogeneity of residential buildings, neighborhoods, and public spaces. Diversity, therefore, not only stimulates creativity and fosters economic growth, but also allows for social mixing. However, a policy that promotes diversity should not transgress people's freedom of choice by forcing them to leave their place of residence under the pretext of producing diversity ([41]: 16). Deliberative and procedural democracy (along the lines of Habermas) is criticized for its excessive faith in the efficacy of open communication, as it ignores the reality of structural inequality and hierarchies of power. Therefore, rather than emphasizing process and participation per se, Fainstein stresses the importance of achieving an adequate representation of interests (content), and an evaluation of the results of urban planning; that is, she underlines the relevance of the criterion of equity, which must be applied to the evaluation of policies, if greater justice is to be achieved ([40]: 35). Equity for Fainstein refers to a distribution of material and non-material benefits derived from public policy that does not favor those who are better placed at the initial starting point, in the Rawlsian sense. It does not require, on the other hand, that each person is treated in the same way, but that the treatment is appropriate ([40]: 36). The application of the criterion of equity in urban decision-making thus elevates the position of weaker and lower income groups in terms of the decisions that are made in particular situations.

These three criteria, described here only in general terms, do not automatically support each other and, indeed, in any particular situation, may conflict. Although Fainstein acknowledges that in policy practice they are difficult to combine, she gives priority to equity. Along these lines, in the conclusion of *The Just City*, she offers a list of specific policies for each.

The goal of a just city has been strongly criticized [56, 69, 74, 75] on at least two grounds: first, because it is seen as a narrow and weak analysis to address the injustices that are inherent in the capitalist system, an unjust system in itself that today extends globally, and, second, because the just city concept that assumes justice is achievable within a capitalist society without the need for structural change in society is inadequate and doomed to failure—it is a myth. Fainstein accepts these criticisms, but maintains that, moving within the capitalist regime of rights and freedoms does not avoid conflict, as continued pressure for justice will gradually change the regime. In this sense, forcing decision-makers to make justice their primary consideration in urban policies would be more than just marginal change ([40]: 6). In her own words: "The hope underlying the just city discussion is that it can shift the rhetoric around urban policy from a focus on competitiveness to a discourse on justice" ([42]: 14).

3.3 The Ethical City

The ethical city project is undoubtedly the most recent and innovative proposal. One of its main promoters is Brendan F. D. Barrett, in collaboration with Ralph Horne and John Fien. According to these authors, the ethical city does not compete with the just city or with other models. Rather, it suggests that there can be both a just and an ethical city. "While the former seeks to determine what is fair, the latter requires that we consider what the right thing to do is. We therefore position the ethical city as a meta-concept or meta-framework" ([5]: 10).

While there are numerous initiatives to promote urban sustainability, smart cities, or healthy cities, their managers claim that less has been documented about cities managed, planned, and operationalized according to an ethical framework. The need for the ethical city is based on the fact that 95% of urban growth up to 2050 will occur in developing countries in Asia and Africa—a process that will be verified through informal settlements. For wealthier cities, economic disparities and the marginalization of large groups of citizens will worsen. This is without exaggerating climate change, resource constraints related to land, air, water, fod and energy, urban traffic, post-industrial restructuring, etc. In this scenario the authors argue for the urgent need to rebalance the obsession with economic growth and profitability with urban development, in order to plan for broader social goals.

What is the ethical city? The authors acknowledge that, to date, the term has rarely been used. At the outset, the ethical city contrasts with what the authors call the "neoliberal city," which expresses free-market utopianism in which local government is based on business, the citizen on the consumer, and governance on business management. In the ethical city, by contrast, local government participates in the creation of the city with its community, citizens have rights and responsibilities, and governance is a collaborative process ([5]: 3). According to the expectation held by Ralph Horne: "The ethical city, in many respects, is the next logical step for human progress in the urban age." It is the way to respond to the most pressing global problems of cities affected by climate change, as well as the various social and economic problems, such as inequality, marginalization, poverty, and lack of opportunities for the most vulnerable majority of the population.

Along these lines, four interrelated dimensions that characterize the ethical city are concretely proposed: ethical leadership/governance, ethical planning, ethical business environment, and ethical citizens ([5]: 5, [6]). They also offer an agenda for the ethical city that draws on the principles of the UN Global Compact in the areas of human rights, labor, environment, and anti-corruption, as well as associated dialogues around the Sustainable Development Goals (SDGs) and the New Urban Agenda (UN-Habitat III) [58].

Among the initiatives that promote an emphasis on the ethical city as an urgent objective of the New Urban Agenda is the Ethical Cities and Urban Innovation Forum Europe, held in Barcelona in 2016, which was a continuation of the Urban Thinkers Campus, "Ethical Cities: Locking in Liveability," held the same year in Melbourne (Australia), an event organized by RMIT University in collaboration with UN Global Compact Cities Programme and the organization World Vision International. These meetings bring together experts from the private sector, academia, public administrations, and various civil society organizations, and explore the ethical city through three central themes: Ethical Urban Development; Resilience; and, Inclusion and the Right to the City.

4 The Ethics of the City as Applied Ethics

In the introduction to this chapter, I pointed out that the challenge of reconstructing the ethical discourse on the city remains. As we have seen, there are at least three responses to this challenge: the *right to the city*, which was born within the contemporary neo-Marxist tradition with Henry Lefebvre; the *just city*, which in Susan Fainstein's proposal follows the paths of the liberal theory of Rawlsian justice; and, finally, the *ethical city*, whose authors assume the programmatic keys of a pragmatic ethics. While recognizing the richness of each of these proposals (above all, the long tradition of the right to the city), only the just city (which, strictly speaking, is a theory of urban justice) dares to offer a philosophically founded response.

In my opinion, this last aspect is crucial, since offering an adequate response not only involves taking charge, reflexively and critically, of the reality and the problems of the so-called *urban question* (for example, denouncing the injustices that are noticed in the process of urbanization dominated by neoliberal capitalism and its individualistic ethics), but also demands a rational search for a normative ethical foundation. Properly understood, the satisfaction of this aspect does not represent the aspiration of a mere "uprooted universalism" ([56]: 40), but rather maintains as a demand of rational validity, at least if it is taken seriously (as Fainstein warns), the overcoming of the positivist scientific moment in the field of social sciences.

Having said this, and entering the debate, in this last section I offer a sketch of the main features of an ethics applied to the city following the methodological perspective of a critical hermeneutics of human activities, as proposed by Cortina ([24–28, 26]: 151–183) and Conill [20]. When assuming this method, we understand that it is the most appropriate to carry out an ethical reflection applied to the city whose main objective does not consist of immediately prescribing concrete actions, but rather it tries to mediate, offering a frame of reference to carry out the deliberation and the concrete decision making ([23]: 121). It is not, therefore, about directly telling the social and affected agents (citizens, professionals, politicians, local and national government, companies, etc.) what should or should not be done in the urban sphere, but rather to offer a compass that orients the action in the concrete situations that are nested in the process of urbanization, or in the collective construction of the city.

From this perspective, an ethics of the city has to try to discover, in the very heart of the city: the goals that give it meaning and social legitimacy; the specific maxims by which it must be oriented to achieve these goals; the virtues that must be cultivated by those who work in this area; the values that must be embodied, and the philosophical foundation of the maxims, which lends them rational validity and provides a criterion for criticism ([25]: 45). A similar itinerary can be posed through the following questions:

1. What are the goals (or internal goods) of the city?
2. What are the philosophically grounded maxims (ethical principles) that guide the pursuit of these goals?
3. What are the virtues to be cultivated by different urban stakeholders (citizens, social groups, local and national government, professionals, etc.)?
4. What are the values to be embodied by the city?

Unravelling an answer to these and other questions that constitute the ethical framework for the city is not a task that engages only the moral philosopher accustomed to working alone; rather, it involves a cooperative interdisciplinary work in which a dialogue is established between the moral philosopher and experts in the urban field such as architects, engineers, sociologists, anthropologists, geographers and urban planners, among others. In this dialogue, of course, citizens, interest groups, local governments and businesses also join in. To a large extent, this cooperative dialogue has already begun to take its first steps. Thus, while awaiting further deepening, we can only venture some tentative and preliminary answers for the time being.

Without the intention of offering a definition of the city, something on which those who have studied the subject have not been able to agree [14, 17], it is convenient to start by unraveling its goal or internal good. In carrying out this inquiry we are inspired by the Aristotelian tradition that conceives human activities as oriented by a *telos* (an end, a goal, a good) that is inherent to praxis and that partially defines it. From this philosophical viewpoint, it can be argued that the city possesses an ethical reason that is rooted in itself, and from which it derives its full meaning and social legitimacy. This last aspect is of utmost importance, since, as Cortina argues: "A social activity must be socially legitimized, and it is only so if society accepts as good the goods it provides and also the means it uses to achieve them" ([25]: 46–47).

According to what has been said, we can understand the city as an authentic *social practice*, as MacIntyre understands the latter following in the wake of Aristotle.[9] From this perspective, the city can be characterized as a cooperative social practice in which a variety of agents intervene. Their purpose is to create and recreate the habitat in which the personal and collective life of human beings, their different forms of association, relations, and exchanges in urban structured spaces (be they social, political, cultural, economic, religious, or ecological) unfolds and grows. From here we can derive a variety of concrete ethical demands that are coherent, for example, with the World Charter for the Right to the City, or with the Rawlsian principle of difference applied to the urban context. These demands are also coherent with one of the fundamental ethical principles of urbanism, according to which one of the goals of any city must be to "guarantee to all inhabitants equal access to goods and services that are common or necessary to all (housing, transport, public space, etc.)" ([12]: 20).

Although this delimitation makes an abstraction of the empirical determinations that have traditionally served to define the city, such as density and extension, or also urban culture, together with Park it serves us to establish that "the city and the urban environment represent for man the most coherent and, in general, the most satisfactory attempt to recreate the world in which he lives according to his own desire" ([80]: 115). In this sense, as Harvey comments, "the question of what kind of city we want cannot be divorced from the question of what kind of people we want to be, what kinds of social relations we seek, what relations to nature

[9] It would be necessary to grant that Lefebvre ([61]: 57, 119–120) already anticipated the category of "social practice" when reflecting on integration, participation and segregation, however, here the idea is different. For MacIntyre ([67]: 187), "practice" is understood to be "any coherent and complex form of socially established cooperative human activity through which goods internal to that form of activity are realized in the course of trying to achieve those standards of excellence which are appropriate to, and partially definitive of, that form of activity, with the result that human powers to achieve excellence, and human conceptions of the ends and goods involved, are systematically extended."

we cherish, what style of life we desire, what aesthetic values we hold" ([55]: 4). At this point, there is no need to push the point on what has already been said about the urban pathologies that currently beset the internal good of cities: poverty, social exclusion, gentrification, hyper-degraded urban areas, congestion, environmental pollution, etc. These and other pathologies represent the corruption of urban space, or what we have called the *tragedy of cities.*

From our understanding of the internal good, we can take another step in the reconstruction of the ethical discourse on the city by considering the ethical principles that constitute the deontological framework that modulates the critical instance towards the means used to achieve the purpose of the city. Without going into the complete process of ethical foundation of the principle of the ethics of discourse—in the line of Apel ([3]: 380 and 381), Habermas ([52]: 117), and Cortina [27, 23], the recognition of every person as a valid interlocutor is presented here, which means that no one should be excluded in the dialogic processes oriented to the establishment of norms. Given that this principle has to be embodied according to the realities of each field of human activity, we could venture that it possesses a decisive importance in the practical discourses that orient the processes of urbanization, for example, in the elaboration of public policies on social housing, in the distribution of public spaces, etc. Strictly speaking, the dialogic principle serves as a normative framework to guide the actual practical discourses in the urban sphere, which are often distorted by goods external to the city, such as the predominance of the purely economic interest imposed by the market.

Parallel to this dialogical principle that guides practical discourse, there is also the principle of human dignity, whereby every human being is to be treated as an end in himself or herself, and not merely as a means. At the present time we can affirm that this principle has become the ethical cornerstone of every city. Demands of justice that every city is obliged to embody stem from this principle such as: respect for equal dignity and the non-instrumentalization of people in the processes of urbanization and urban renewal; the construction of urban spaces that contribute to empowerment, to the development of people's capacities, and not to their degradation; the equitable distribution of urban burdens and benefits over the speculative exchange value of the price of land; the responsibility for present and future generations, together with the care of the environment, which today represent demands of sustainable urban development ([29]: chap. 9).

5 The Urban Revolution Will Be Ethical or It Will Not Be

As I argued at the beginning of this chapter, the ethics of the city constitutes a study and research project that has just begun to take its first steps. What has been said so far represents only a sketch of some of its dimensions. The data on the human and environmental consequences that make the process of urbanization an unsustainable reality make it a challenge of the first magnitude socially, politically, economically, ecologically—and, of course, ethically. Despite the abundant declarations by governments and international organizations, and the multiple initiatives of numerous civil society movements at the local, regional, and global levels, it is undeniable that, at least until now, urban development appears as a process dominated unilaterally by the commodification of the city. Everything seems to support Lefebvre's diagnosis of the substitution of use value for exchange value, citizenship for the client, public space for the real estate business. The projections about the city and its problems for this century are overwhelming; however, there remains the conviction and the utopian-realist hope that another city is still possible.

Following Lefebvre, David Harvey has argued that "the revolution in our times has to be urban-or-nothing" ([55]: 25). For my part, and from the perspective developed in this brief chapter, I would like to add that, if such an urban revolution is to succeed, it has to be founded on ethics—or it will not *be*. This is so, because any revolution, urban or not, if it is not ethically grounded, will sooner rather than later end up succumbing to the totalitarian demands of political bureaucracy and economic efficiency. Certainly, appealing to the ethical rationality of the city alone does not represent the solution to urban pathologies. Ethics isolated from economic and political systems is a mere abstraction. Consequently, it must necessarily be embodied in economic and political institutions, which must be placed at the service of the city's internal goods.

References

1. Álvarez AMª (2013) (Des)Igualdad socio espacial y justicia espacial: nociones clave para una lectura crítica de la ciudad. Polis 12(36):265–287
2. Ansay P, Schoonbrodt R (eds) (1989) Penser la ville: choix de textes philosophiques. Archives d'Architecture Moderne, Brussels
3. Apel K-O (1985) La transformación de la filosofía, vol II. Taurus, Madrid
4. Arrango J (1970) The urbanization of the Earth. Beacon Press, Boston
5. Barrett B, Horne R, Fien J (2016) The ethical city: a rationale for an urgent new urban agenda. Sustainability 8(11):1–14
6. Barrett B (2015) Ethical cities are the future, our world. http://ourworld.unu.edu/en/ethical-cities-are-the-future
7. Bassols M, Donoso R, Massolo A, Méndez A (Comp.) (1988) Antología de Sociología Urbana. Universidad Nacional Autónoma de México, Mexico
8. Bettin G (1982) Los sociólogos de la ciudad. Editorial Gustavo Gili, Barcelona
9. Borja J (2003) La ciudad conquistada. Alianza Editorial, Madrid
10. Borja J (2007) Para reconstruir un discurso ético sobre la ciudad. Interview by S. López to Jordi Borja. El Viejo Topo 231:51–55
11. Borja J (2013) Revolución urbana y derechos ciudadanos. Alianza Editorial, Madrid
12. Borja J, Carrión F (2017) Introducción. In: Borja J, Carrión F, Corti M (eds) Ciudades resistentes, ciudades posibles. Editorial UOC, Barcelona, pp 17–58
13. Cacciari M (2010) La ciudad. Editorial Gustavo Gili, Barcelona
14. Capel H (1975) La definición de lo urbano. Estudios Geográficos 138–139:265–301
15. Carreras C, MorcuendeA (2016) Marxismos y utopía. Viejas y nuevas propuestas urbanas. http://www.ub.edu/geocrit/xiv-coloquio/CarrerasMorcuende.pdf. Accessed 5 June 2017

16. Castells M (1977) The urban question. A Marxist approach (translated by Alan Sheridan). Edward Arnold, London
17. Castro P et al (2003) ¿Qué es la ciudad? Aportaciones para su definición desde la prehistoria. Scripta Nova. Revista Electrónica de Geografía y Ciencias Sociales. http://www.ub.edu/geocrit/sn/sn-146(010).htm
18. Chan JKH (2019) Urban ethics in the Anthropocene. Palgrave Macmillan, Singapore
19. Colom F (2014) Max Weber y la ciudad. In: Morcillo A, Weisz E (eds) Max Weber: una mirada iberoamericana. FCE, Mexico City, pp 363–390
20. Conill J (2006) Ética hermenéutica. Tecnos, Madrid
21. Correa Casanova M, Montoya J, Mealla EP (2019) Ética Aplicada. Perspectivas desde Latinoamérica. Uniandes, Bogotá
22. Correa Casanova M (2019) Hacia una ética de la ciudad. In: González E, Siurana JC, López JL, García-Granero M (coords.) Ética y democracia: desde la razón cordial. Editorial Comares, Granada, pp 251–257
23. Cortina A (1996) El estatuto de la ética aplicada. Hermenéutica crítica de las actividades humanas. Isegoría 13:119–134
24. Cortina A (2002) La dimensión pública de las éticas aplicadas. Revista Iberoamericana de Educación 29:45–64
25. Cortina A (2010) Justicia cordial. Trotta, Madrid
26. Cortina A, Martínez E (1996) Ética. Akal, Madrid
27. Cortina A (1985) Razón comunicativa y responsabilidad solidaria. Ética y política en in K.-O. Apel. Sígueme, Salamanca
28. Cortina A (2003) El quehacer público de las éticas aplicadas: ética cívica transnacional. In: Cortina A, García Marzá D (eds) Razón pública y éticas aplicadas. Tecnos, Madrid, pp 13–42
29. Cortina A (2009) Ética de la razón cordial. Ediciones Nobel, Oviedo
30. Costes L (2009) Lire Henri Lefebvre. Le droit a la ville: Vers la sociologie de l'urbain. Ellipses, Paris
31. Costes L (2010) Le Droit à la ville de Henri Lefebvre: quel héritage politique et scientifique? Espaces et sociétés 1(140–141):177–191
32. Costes L (2011) Del 'derecho a la ciudad' de Henri Lefebvre a la universalidad de la urbanización moderna. Urban 2:89–100
33. Davis M (2006) Planet of slums. Verso, London and New York
34. Davis JC (1985) Utopía y la ciudad ideal. F.C.E., Mexico
35. De Mattos CA, Link F (eds) (2015) Lefebvre revisitado: capitalismo, vida cotidiana y el derecho a la ciudad. RIL editors, Santiago
36. Deneulin S (2014) Crear ciudades más justas para la vida: una combinación del derecho a la ciudad y el enfoque de las capacidades. In: Suarez AL, Mitchell A, Lépore E (eds) Las villas de la ciudad de Buenos Aires: territorios frágiles de inclusión social. Educa, Buenos Aires, pp 253–281
37. Ege M, Moser J (eds) (2021) Urban ethics. Conflicts over the good and proper life in cities. Routledge, London and New York
38. Engels F (1974) El problema de la vivienda y las grandes ciudades. Gustavo Gili, Barcelona
39. Engels F (1984) La situación de la clase obrera en Inglaterra. Ediciones de Cultura Popular, Mexico
40. Fainstein S (2010) The just city. Cornell University Press, Ithaca and London
41. Fainstein S (2013) Planning, justice and the city. Urban 6:7–20
42. Fainstein S (2014) The just city. Int J Urban Sci 18(1):1–18
43. Fainstein S (2009) Planning and the just city. In: Marcuse P, Connolly J, Novy J, Olivo I, Potter C, Steil J (eds) Searching for the just city. Routledge, London, pp 19–39
44. Fainstein S (2009) Spatial justice and planning. Justice spatiale | spatial justice, no. 1. http://www.jssj.org
45. Fraser N (1997) Justice interruptus. Routledge, New York

46. Fraser N (2003) Social justice in the age of identity politics. In: Fraser N, Honneth A (eds) Redistribution or recognition? A political-philosophical exchange. Verso, New York, pp 7–109
47. García Vásquez C (2016) Teorías e historia de la ciudad contemporánea. Editorial GG, Barcelona
48. Garnier J-P (2012) El derecho a la ciudad desde Henri Lefebvre a David Harvey. Entre teorizaciones y realización. Ciudades 15(1):217–225
49. Gasic I, Narváez A, Quiroz R (comps.) (2015) Reapropiaciones de Henri Lefebvre: crítica, espacio y sociedad Urbana. Editorial Triángulo, Santiago de Chile
50. Gilson E (1965) The metamorphoses of the city of God. Rialp, Madrid
51. Goonewardena K, Kipfer S, Milgrom R, Schmid C (eds) (2008) Space, difference, everiday life. Reading Henri Lefebvre. Routledge, New York and London
52. Habermas J (1985) Conciencia moral y acción comunicativa. Península, Barcelona
53. Habermas J (1986–89) Theory of communicative action (translated by Thomas McCarthy). Polity Press, Cambridge
54. Harvey D (1993) Social justice and the city. Blackwell, Oxford
55. Harvey D (2012) Rebel cities. From the right to the city to the urban revolution. Verso, London and New York
56. Harvey D, Potter C (2009) The right to the just city. In: Marcuse P, Connolly J, Novy J, Olivo I, Potter C, Steil J (eds) Searching for the just city. Routledge, New York, pp 40–51
57. Heidegger M (1997) Construir, habitar, pensar. Filosofía, ciencia y técnica. Editorial Universitaria, Santiago de Chile, pp 197–219
58. Horne R, Barrett B, Das J (2016) Towards the ethical city. In: Briefing paper for the 'ethical cities-locking in liveabity' Urban thinkers campus. RMIT University, Melbourne
59. Hénaff M (2014) La ville qui vient. Mille at une Nuits, Paris
60. Jacobs J (1992) The death and life of great American cities. Vintage Books, New York
61. Lefebvre H (1969) El derecho a la ciudad. Península, Barcelona
62. Lefebvre H (1970) La révolution urbaine. Gallimard, Paris
63. Lefebvre H (1974) La production de l'espace. Éditions Anthropos, Paris
64. Lefebvre H (1978) De lo rural a lo urbano. Ediciones Península, Barcelona
65. Lefebvre H (1976) Espacio y política. El derecho a la ciudad, vol II. Península, Barcelona
66. Lin J, Mele C (eds) (2013) The urban sociology reader. Routledge, London and New York
67. MacIntyre A (2007) After virtue. A study in moral theory (3rd edn). University of Notre Dame Press, Notre Dame, IN
68. Marcuse P, Connolly J, Novy J, Olivo I, Potter C, Steil J (eds) (2009) Searching for the just city. Routledge, New York
69. Marcuse P (2009) Spatial justice: derivative but causal of social justice. Justice spatiale | spatial justice, no. 1. http://www.jssj.org
70. Mathivet C (coord. and ed) (2016) Develando el derecho a la ciudad. Passarelle Collection, Coredem, Ritimo, HIC-AL, no 15. http://www.coredem.info/IMG/pdf/passarelle-droitville-es-okimpr.pdf
71. Meagher SM (2008) Philosophy and the city: classic to contemporary writings. State University of New York, Albany
72. Mostafavi M (ed) (2017) Ethics of the urban. The city and the spaces of the political. Lars Müller Publishers, Zürich
73. Mumford L (1961) The city in history. Harcourt, Brace & World, New York
74. Musset A (2015) El mito de la ciudad justa. Bitácora 25(1):125–139
75. Musset A (2016) De la inclusión a la resiliencia: Las palabras mágicas de la 'ciudad justa'. In: Mathivet C (coord. and ed) Develando el derecho a la ciudad. Passarelle Collection, Coredem, Ritimo, HIC-AL, No. 15, pp 50–59. http://www.coredem.info/IMG/pdf/passarelle-droitville-es-okimpr.pdf.
76. Nancy J-L (1999) La ville au loin. Mille et Une Nuits, Paris

77. Nehls N, Ortíz E, Zarate ML (comp. and eds) (2008) El derecho a la ciudad en el mundo. Compilación de documentos relevantes para el debate. Habitat International Coalition-Regional Office for Latin America, Mexico. http://www.hic-al.org/noticias.cfm?noticia=640
78. Nussbaum M (2000) Women and human development. Cambridge University Press, Cambridge
79. Nussbaum M (2006) Frontiers of justice. Harvard University Press, Cambridge
80. Park RE (1999) La ciudad y otros ensayos de ecología urbana. Ediciones del Serbal, Barcelona
81. Park RE (1992) The city: suggestions for the investigation of human behavior in the urban environment. In: Park RE, Burgess EW, McKenzie RD (eds) The city. The University of Chicago Press, Chicago and London, pp 1–46
82. Pereira P (org.) (2008) A Filosofia e a Cidade. Campo das Letras, Porto
83. Pereira P (org.) (2011) A Filosofia e a Cidade, vol II. Edições Afrontamento, Porto
84. Rawls J (1971) A theory of justice. Harvard University Press, Oxford
85. Rojas A (2012) La ciudad inhóspita promovida por Heidegger. Isegoría 47:499–517
86. Sayer A, Storper M (1997) Ethics unbound: for a normative turn in social theory. Environ Plan D-Soc Space 15(1):1–17
87. Sennett R (2018) Building and dwelling. Ethics for the city. Farrar, Straus and Giroux, New York
88. Soja EW (2000) Postmetropolis. Critical studies of cities and regions. Blackwell, Oxford and Malden, MA
89. Soja E, Kanai M (2008) The urbanization of the world. In: Burdett R, Sudjic D (eds) The endless city: the urban age project by the London School of Economics and Deutsche Bank's Alfred Herrhausen Society. Phaidon, London, pp 54–69
90. Strauss L (2006) La ciudad y el hombre. Katz, Buenos Aires
91. Sugranyes A, Mathivet C (eds) (2010) Cities for all: proposals and experiences towards the right to the city. http://www.citiesalliance.org/sites/citiesalliance.org/files/Cities_For_All_ENG.pdf
92. UN-Habitat (1996) An urbanizing world. Global report on human settlements 1996. Oxford University Press, Oxford
93. UN-Habitat (2003) The challenge of slums. Global report on human settlements 2003. London
94. UN-Habitat (2016) Urbanization and development: emerging futures. World Cities Report 2016. Nairobi
95. Wacquant L (2008) Urban outcasts: a comparative sociology of advanced marginality. Polity Press, Cambridge
96. Weber M (1987) La ciudad. Ediciones de La Piqueta, Madrid
97. Weber M (2002) Economía y sociedad. F.C.E., Mexico
98. Young IM (1990) Justice and the politics of difference. Princeton University Press, Princeton
99. Young IM (2000) Inclusion and democracy. Oxford University Press, Oxford
100. Zukin S (2006) David Harvey on cities. In: Castree N, Gregory D (eds) David Harvey. A critical reader. Blackwell Publishing, Malden, MA, pp 102–120

Ethics of Humor

10

Juan Carlos Siurana Aparisi

Abstract

The bibliography on humor is very extensive, but there are few publications that have tried to elaborate an ethics of humor. I will begin by recounting the basic ideas of Ronald de Sousa, who tells us that our laughter reflects our moral values. Three theories of the causes of the comic have been defended throughout history: the superiority theory, the incongruity theory, and the tension release theory. In connection with those theories, I will show the important proposals for the ethics of humor by Francis H. Buckley and John Morreall. Having explained these authors, I will go on to offer my own proposal for the ethics of humor, as a new ethical theory and as a new applied ethics. I will show various fields of its application, highlighting its contribution to ethical improvement in health care, business and education.

Keywords

Humor • Laughter • Moral values • Ethics of humor • Humor in health care • Humor in business • Humor in education

This study is part of the Scientific Research and Development Projects "Ethics and Democracy in the face of the challenges of Artificial Intelligence" PID2019-109078RB-C22 and "Cordial ethics and inclusive democracy in a technologized society", PID2022-139000OB-C21, funded by MCIN/AEI /https://doi.org/10.13039/50110001103310.13039/501100011033, and in the activities of the Research Group of Excellence of the Prometheus 22 Program CIPROM/2021/072 of the Generalitat Valenciana.

J. C. Siurana Aparisi (✉)
Universidad de Valencia, Valencia, Spain
e-mail: Juan.C.Siurana@uv.es

A. Cortina and M. Correa Casanova (eds.), *Applied Ethics from Medicine to Humor*, Integrated Science 21, https://doi.org/10.1007/978-3-031-59009-2_10

1 Introduction

The bibliography on humor is very extensive. We can gather several thousand contributions from different academic disciplines, both throughout history and nowadays, but there are really very few publications that have tried to elaborate an ethics of humor as applied ethics.

There are, of course, books by classical authors who have addressed the topic of humor from an ethical perspective, including Aristotle [3], Aquinas [2], Shaftesbury [52], Kant [33], Kierkegaard [34], Nietzsche [44], Lipps [37], Höffding [28], Bergson [6], Freud [22, 23], or Pirandello [47], although they have mostly done so by offering small comments on it in works devoted to other interests (only in some cases, in works devoted entirely to humor). We also find contemporary authors who have written an article in a scientific journal, or a book chapter on the "ethics of humor," within a book that pursued other general interests. But there is practically no attempt to deal in a monographic way with the ethics of humor as applied ethics in the current sense of the term.

In recent decades, interest in the study of humor has been reinforced by the creation of several international scientific societies, most notably the International Society for Humor Studies, which held its first conference in Cardiff, Wales, in 1976. This society publishes the quarterly *Humor: International Journal of Humor Research*, which has become an interdisciplinary forum for the publication of high-quality articles on humor as an important and universal human faculty.

To explain here what I call the ethics of humor, I will begin by recounting the basic ideas of a very interesting current author, Ronald de Sousa, who tells us that our laughter reflects our moral values [18].

Three theories of the causes of the comic have been defended throughout history: the superiority theory, the incongruity theory, and the tension release theory. In connection with those theories, in this chapter I will show the proposal for the ethics of humor by two prominent authors: Francis H. Buckley, in his book entitled *The Morality of Laughter* [11], and John Morreall in his book entitled *Comic Relief. A Comprehensive Philosophy of Humor* [43]. Buckley places himself within the framework of the theory of superiority and Morreall within the theory of incongruity. This position permeates, in a decisive way, their ethical proposals.

Having explained these authors, I will go on to offer my own proposal for the ethics of humor, which I have set out in several works and, systematically, in my book *Ética del humor. Fundamentos y aplicaciones de una nueva teoría ética* [59].

In the final part of the chapter I will show various fields of application of the ethics of humor, highlighting its contribution to ethical improvement in health care, business and education.

I end the chapter by offering some brief conclusions on the above.

2 The Definition of "Humor"

The first thing we need to do when talking about humor is to agree on how we define it. My definition of humor uses several related concepts that in turn also require definition. Therefore, in what follows, and drawing mainly from the voluminous work by Martin [39], I will offer my definitions of several concepts starting with the one that primarily concerns us:

a. *Humor*: The capacity to perceive something as funny, which activates the emotion of mirth, expressed through smiling or laughing.
b. *Mirth*: A pleasurable emotion related to joy that is expressed by smiling or laughing, provoked mainly by something funny that humor is able to perceive, although it can be provoked by other causes, such as tickling.
c. *Smile*: Physical expression of mirth of low intensity (arching of the lips).
d. *Laughter*: Physical expression of mirth of high intensity (moderate laughter) or very high intensity (loud laughter).
e. *Funny or comical*: A quality of living beings and things that, when captured, provokes the emotion of mirth. That which possesses this quality is called funny or comical.

So, it is convenient to distinguish between "humor," which is a capacity, and "mirth," which is an emotion that refers to the joy that is expressed through laughter.

3 The Causes of Something Being Funny

Almost all philosophers, when dealing with the subject of humor, have focused on trying to find out the causes of the comic, that is, to know why something makes us laugh.

Scholars on the philosophy of humor agree that, throughout the history of philosophy, there have been at least three important theories on the causes of the comic:

1. *Superiority theory*. This is the oldest of them all. For this theory we judge a situation comical when we see that the other fails while we think that we would overcome such a situation with more probabilities. We laugh at the shortcomings of others. In that sense, laughter expresses a feeling of superiority. Among its main representatives would be Plato [48, 49] and Hobbes [27].
2. *Incongruity theory*. For this theory we laugh when what happens undergoes an unexpected and incongruous change, which surprises us suddenly. Humor here means distortion of the function, going out of the norm. It would have been defended by authors such as Kant [33] and Kierkegaard [34].

3. *Theory of the release of tension.* This theory understands that the human being acts under effort, struggle and tension, and this excites the nervous system. Laughter provides relief for nervous or psychic tension, ensuring a restoration of equilibrium. This theory has been proposed, prominently, by Freud [22, 23].

I have found the subject of the causes of the comic very interesting, but in my work I have focused on investigating: (1) whether humor is a sufficiently relevant phenomenon on which to base a new ethical theory; and, (2) its application to the multiple situations of life. The first has led me to defend the ethics of humor as an ethical theory; the second, to explain what the ethics of humor as applied ethics consists of. I will show these two dimensions briefly below, emphasizing above all its dimension as applied ethics.

4 The Ethical Evaluation of the Emotion Expressed by Laughter

Ronald de Sousa has elaborated a proposal according to which "emotions can be rationally evaluated" ([18]: 227), within this framework he ethically evaluates laughter.

Laughter, as this author understands it, is a response to what is amusing, funny, comical, or ridiculous about formal objects. He brings it all together under the name of *the Funny.*

He is interested in the Funny, in our emotional response to it, and in the possibility of rationally—in this case, ethically—evaluating that response.

To the extent that we think that it is sometimes irrational or morally wrong to laugh, it is worth trying to rationalize the principles behind such assessments in order to construct what De Sousa calls an "Ethics of Laughter."

Thus, his work on the ethics of laughter, which is, above all, an ethics of mirth, can be understood within the framework of a more general discussion of the role of emotions in moral life.

According to him, reason sometimes tells us that it is not right to laugh, limiting the emotion of mirth or, at least, the expression of it through laughter; for example, we are all able to perceive the difference between laughing at someone and laughing with someone. In many cases, reason will limit our emotion to prevent us from laughing at someone.

According to Ronald de Sousa, an ethics of laughter, based on common sense, would defend the following:

> Laugh when it's funny, grow up and stop snickering at dirty jokes, don't laugh at cripples (unless you are one yourself), and *show respect.* To show respect means not to laugh, snicker, titter, chortle, giggle, or even chuckle when it's Too Sad, when it would be Unkind to, when it would Offend a Sacred Memory, and when it might be taken to Insult a Mother, a Country, or a Religion. ([18]: 228)

But a few precepts do not constitute an ethic. The author wonders whether anything can properly be called an ethics of laughter and answers yes, insofar as laughter is a reflection of a person's character.

There are cases where we can say, "If you are able to laugh at something like that, you must be insensitive, rude, or cruel".

For the Kantian, what is important is the intention of laughter. But the fact that laughter arises almost without our noticing it—as a trait of our character, as something natural integrated into our personality—makes it inappropriate for us to refer to the intentions of laughter. We don't laugh with an intention, we laugh as a consequence of our character.

According to De Sousa, if we can ask "what were you trying to achieve by laughing," then there was no genuine laughter. Genuine laughter has no intentionality, we laugh at what we find funny, at what we find comical. And it depends on our character whether laughter arises or not. So, laughing is a symptom of something. It shows what are the traits of our personality, of our character.

De Sousa focuses on investigating a type of laughter that is susceptible to moral condemnation: malicious laughter, that laughter that is directed especially against our enemies and presupposes that those who laugh share our same animosity towards them.

The key here is to find out if this laughter towards the supposed "enemies" is justified or if it is, in fact, unfair.

When one tells a sexist or racist joke, the "enemies" are women or people of other races, and the assumption is that the listener who hears and laughs at the joke shares those same attitudes toward those groups. And the interesting thing for De Sousa is that we cannot find something funny just by imagining that we share the presupposed attitudes; we must share them internally. He argues that you can't think with ethically incorrect values if you don't have them.

De Sousa affirms that laughing at a malicious joke reveals that the same attitudes are shared, since "attitudes are beliefs that one cannot adopt hypothetically" ([18]: 241).

But those beliefs may be ethically wrong. If laughter is wrong, it is because the emotional commitment is wrong. "The 'unethical'... involves a wrong *assessment of reality*" ([18]: 244). Thus, "laughter is wrong when it is grounded in the deception of self or others" ([18]: 244).

We could say that where the attitude presupposed by a joke is morally questionable, then telling the joke or laughing at it is also morally questionable.

5 Laughter Teaches Us a Life Plan Immune to Ridicule

F. H. Buckley's book *The Morality of Laughter*, first published in 2003, is one of the few books written in recent decades devoted exclusively to thinking about laughter from an ethical perspective [11]. It is, therefore, a crucial book for current thinking about the possibility of constructing an ethics of humor.

For Buckley, one of the strongest cultural cues for how to live come from laughter, the sound of which we can never ignore. We can endure poverty, illness, even shame, but not ridicule, and he adds: "the strategies we employ to immunize ourselves from it constitute a thick code of conduct that I call the morality of laughter" ([11]: preface X).

There are religious, legal, and moral codes that condemn the infliction of harm on others, but the morality of laughter is different, in that it *teaches individuals how to extract enjoyment from their own lives*. He considers that Nietzsche recognized this difference in defending the Gay Science.

Buckley understands that laughter announces and reinforces a code of conduct through the joker's sign of superiority over the person who is the target of his or her jokes. Thus, Buckley defends two theses:

First, the *positive thesis of superiority*: laughter expresses superiority.

According to this author: "There is no laughter without a butt, and no butt without a message about a risible inferiority" ([11]: preface XI). Buckley calls this statement the "*positive* thesis of superiority" and presents it in part I of his book.

With this, the author places himself in a theory of humor that comes from Plato and Hobbes and, up to that point, he does not contribute anything particularly new. But his second thesis, which reads as follows, is original:

Second, the *thesis of normative superiority*: the superiority expressed by laughter is usually true.

The *normative* superiority thesis, which he examines in part II of his book, makes an even stronger claim that "our laughter communicates true superiority and that the target is truly an inferior person" ([11]: preface XI). That is, Buckley thinks that when we laugh at someone we usually have reason to do so.

Buckley himself acknowledges that the first thesis is more plausible than the second, but he stands by his defense of both.

He acknowledges that those who laugh are sometimes inferior brutes or even evil, and that in such cases we should not accept the message they are trying to convey to us with their laughter, but that should not discount the claim that laughter offers valuable lessons about how to live. It reminds us that few emotions are as strong as the fear of being the target of laughter.

Buckley himself considers the possible criticisms of those who think it is possible to laugh without expressing superiority, for example, when we make puns or when one tells a joke to oneself. But he argues that "a signal of superiority may always be found in every honest laugh, or at least in one that proceeds from a mental stimulus and not a merely physical cause, such as tickling or inhaling nitrous oxide" ([11]: preface XI).

If the *positive* thesis were rejected we would have to abandon the *normative* thesis, but he considers that indeed when we laugh we express superiority. We cannot laugh at someone who is the target of our laughter if we feel inferior to him or her.

Regarding the normative thesis, he recognizes that sometimes laughter signals a false superiority, but that leads us to the need to discriminate between different laughter.

He defends what he calls a "mild version of the normative thesis," according to which the general effect of laughter is benign. By pointing out comic vices, laughter teaches us a higher plan for life that is immune to ridicule.

By identifying a target, the message of laughter is both positive and negative. It tells us what to do and what not to do if we want to immunize ourselves against laughter. In doing so, laughter identifies comic virtues and comic vices.

According to Buckley, "comic virtues teach us how to extract joy from life, and their message is necessarily complex" ([11]: preface XII). This author also wonders why the rules of laughter can err, as he tells us that "the impulse to laugh might arise through base motives, factual mistakes, or errors of judgment" ([11]: preface XII).

If virtue is a middle ground, it is often unclear where that middle ground lies, and what constitutes its excess and defect. However, it seems clear that vile laughter would be a *comic vice by excess* (one who laughs when he should not), and that not laughing would be a *comic vice by defect* (one who does not laugh when he should).

Buckley describes a side effect of laughter, which he calls *sociability*. By creating communities with common targets, laughter serves as a bonding tool between those who make jokes and those who hear them. The listener who laughs takes the side of the joker against the target, and cements a relationship of trust between the listener and the joker.

6 Humor Promotes Ethical Virtues

A different perspective on the ethics of humor is presented by John Morreall, especially in his book *Comic Relief. A Comprehensive Philosophy of Humor* ([43]: 90–124), where he places himself in the framework of the theory of humor as incongruity.

In his book he distinguishes between two types of *humor ethics*: the *negative*, which would defend the importance of limiting humor because it promotes vices; and the *positive*, which describes the moral virtues that can be fostered through humor. In the following, we will leave aside the vices attributed to humor and show which virtues it can promote.

Although the vast majority of religious and philosophical statements about laughter have associated it with vice, some have associated it with positive states, such as mirth or joy. In the New Testament, for example, Jesus seems to regard laughter in this positive sense when he states, "Blessed are you who weep now, for you will laugh" ([29]: Luke 6:21).

Another positive sense of thinking about humor has been associated with play. According to Aristotle, in the life of a human being it is important to make room for leisure and fun. Some people take fun to excess, and Aristotle calls them "vulgar buffoons." But between buffoonery and complete seriousness, there is a middle ground in which lies the virtue that leads to happiness, practicing humor at the right time and place, and to the right degree. This *eutrapelia*, or gentle jest, this special kind of witty disposition, counts as a virtue in his *Nicomachean Ethics* [3].

Aristotle's comments on eutrapelia are not taken into consideration until Thomas Aquinas in medieval times adapts them to Christian scholastic philosophy: in his *Summa Theologiae* he discusses humor and play, and his position is similar to Aristotle's. Human beings need occasional breaks from serious activity. Humor and other forms of play offer us that rest.

The person who has the moral virtue associated with playfulness and humor, Thomas Aquinas calls an *eutrapelle*, a pleasant person with a happy chaste mind who gives his or her words and deeds a cheerful turn (Aquinas [2]: question 168).

Morreall adheres to the proposition of Aristotle and Thomas Aquinas that humor can be virtuous. Both thinkers understand virtue as a kind of excellence of the person, and distinguish between intellectual and ethical virtues. Morreall argues that humor can foster both types of virtue. We will highlight, below, his contribution to the ethical virtues.

According to Morreall, humor promotes the following ethical virtues ([43]: 90–124):

1. *Patience.* Looking at things with a sense of humor, we don't expect them to happen as fast as we would like.
2. *Tolerance.* Humor correlates with open-mindedness and, in social interactions, a willingness to see things differently makes us better understand other people, what they think and how they act. In this way, humor can help reduce social friction. Sammy Basu has examined how humor promotes religious tolerance ([4]: 147–172).

   One example is the man whose father was Jewish but whose mother was not, and who was not allowed to be a member of a golf club during the time of Nazism. He asked to speak to the president of the club to ask him a question. The question was: "Since I am only half-Jewish, can I belong to the club if I only play nine holes?" The president began to laugh and admitted him to the club.
3. *Kindness.* Saying thank you allows others to relax and not feel threatened. A person who is corrected with grace is more likely to hear the message and act on it. A counter example is when someone threatens others with legal action when they feel unfairly treated. This makes people defensive, unreasonable, and even hostile.
4. *Humility.* When one laughs at oneself, one can defuse conflicting situations and is appreciated by others.

5. *Perseverance.* When humility is combined with patience, when we see ourselves from a comical perspective, we are less likely to be overcome by frustration. Thomas Edison tried over ten thousand combinations of materials before he invented the light bulb. When asked if he felt sad about all those failed attempts he replied, "No, I simply learned a thousand ways of how not to make a light bulb."
6. *Courage.* Occurs when perseverance operates in dangerous situations. Some obvious examples occurred during the Holocaust. As Steve Lipman recounts in his book *Laughter in Hell* [36], for example, in the ghettos, Hitler's main work was not known by its real title *Mein Kampf* (My Struggle) but by *Mein Krampf* (My Cramp).

Morreall concludes that humor promotes many virtues, and those who lack this kind of virtuous humor are at serious risk of being mere pedants.

Having applauded humor for its association with many virtues, Morreall adds that this link is not universal, however. There is no necessary connection between humor and virtues, any more than there is between humor and vices. There is humor that implies selfishness, intolerance and cruelty. Humor is essentially neither virtuous nor vicious.

From this statement we can extract that humor that promotes virtues is a specific type of humor, which in this chapter we are calling "ethical humor."

Other important authors have made interesting contributions to the ethics of humor, which, for reasons of space, I cannot include here, such as Critchley [16] and Hösle [30]. In the following, I will point out what constitutes my proposal for the ethics of humor.

7 My Work on the Ethics of Humor

The ethics of humor has been a topic to which I have devoted much of my research in the last decade. In 2012, I published my first article reviewing the classic and updated literature on the connection between philosophy and humor [53]. That same year I presented what could be the foundations for an ethics of humor [54], and even its application to business [55]. During the following years, I continued to offer results, such as the connection between humor, neurosciences and the end of life [56], the features that the ethics of humor would have from the thinking of some contemporary authors who had addressed the subject [57], and its connection with cultural diversity [58] or with Martha Nussbaum's capabilities approach [60]. But it is really with the publication of my book entitled *Ética del humor. Fundamentos y aplicaciones de una nueva teoría ética* (Ethics of humor: Foundations and applications of a new ethical theory) [59], that I have been able to offer in a broad and articulated way, my vision of the ethics of humor, which, in my proposal, is both an ethical theory and an applied ethics.

In the following, I will show the fundamental contents of my proposal, and condense many of the ideas expressed in the book in connection with the works on this subject by other authors.

8 The Ethics of Humor as Ethical Theory

8.1 The Ethics of Humor and the Paradigm of Biology

I argue that, at present, we are moving from the paradigm of language, which has been prevailing in philosophy for the last hundred years [1], to what I call the "paradigm of biology." This shift is due to the central place that the health and biological sciences are occupying in our philosophical conceptions. Never before has technology so radically threatened the quality of life and the very survival of human beings on earth as in our time [50]; and never before have we achieved such detailed knowledge of our own biology—for example, about our brain—that we can hope to find in it the basis of our moral conduct. Bioethics, ecological ethics, animal ethics, gene-ethics, neuroethics, nanoethics, or the ethics of human enhancement are some examples of major branches of ethics sensitive to this paradigm shift within philosophy.

8.2 Biological Rationality Pursues Health and Survival in Comfortable Conditions

The ethics of humor, as an ethical theory, is fully situated in this new paradigm of philosophy. I argue that our biology contains a rationality, which I call "biological rationality" ([59]: 183–187), which pursues a purpose. That purpose cannot be known with certainty, since its very existence is a metaphysical assertion. We can, however, put a meaningful order to what happens in the world of living beings if we start from the assumption that this purpose is health and survival in comfortable conditions. I also argue that humor is one of the means through which this biological rationality is expressed.

8.3 Discursive Rationality Critically Interprets the Messages of Biological Rationality Transmitted Through Humor

Life continuously transmits us messages whose understanding we approach with our discursive rationality, which is, at the same time, an interpretative and critical rationality. We can never know exactly what biological rationality really intends. But presupposing that it seeks health and survival in comfortable conditions gives meaning to our relationship with the biological world. My theory proposes a dialogue with our biology, and by imagining the very possibility of that dialogue it anticipates, as a condition of the possibility of meaning in that communication, that biology transmits meaningful messages to us.

These meaningful messages constitute biological rationality. And I argue that humor is one of the most interesting mechanisms through which biology transmits meaningful messages to us.

To try to understand them, we need to turn to what I call the "discursive rationality of the ethics of humor" ([59]: 165–181). It is a specific type of discursive rationality, which so far has not been studied in detail, that tries to critically interpret the messages that our biological rationality transmits to us through smiles and laughter.

The ethics of humor unites the biological rationality with the discursive rationality of the ethics of humor, but always recognizes that the biological rationality is the primordial one, from which the discursive rationality emerges, without it being said that it is clearer in its message because it is primordial. The discursive ethics of the ethics of humor would try to bring some clarity to that message, with the limits that our human language has. For example, it is a limit of our language to say that biological rationality seeks survival in comfortable conditions, when it could equally be said that it seeks a life worth living, or a life of quality. What is important is that a meaningful communication with the biological dimension can be established.

8.4 The Clinical-Ethical Method as a Medical-Philosophical Method of the Ethics of Humor

For the interpretation of the messages of biology, the ethics of humor resorts to its own method, which necessarily requires the joint work of philosophers with, at least, professionals in health and biological sciences. I have called the method that I propose the "clinical-ethical method" ([59]: 115–129). It consists of interpreting the signs and symptoms that we perceive in human beings, which can be indicators of what I call "diseases due to a lack of ethics." These diseases have a double component: on the one hand, physical and/or mental pathologies, and on the other, lack of ethics or immoral actions.

I argue that smiling and laughter are *signs* of both health (positive laughter) and illness (negative laughter). The diseases shown by negative laughter have a double component (physical-mental pathology and ethical pathology). And what a person says about their relationship with humor also reveals *symptoms* of their health and ethics. The important thing for the ethics of humor, therefore, is to know how to interpret these signs and symptoms.

8.5 Humor as an Archimedean Point of a New Ethical Theory

I propose *a new ethical theory based on humor* [59], that places humor as the Archimedean point from which we can advance our own understanding as moral subjects, obtain reasons for our moral actions, and show ways to improve ethically—both individually and for society as a whole.

I defend that the capacity to perceive something as funny, characteristic of humor, can be more or less correct from an ethical point of view, and that it is important to educate in *ethical humor*. *Ethical humor* is that which knows how to laugh at the right time, in the right way and to the right degree, and for the right reasons.

But by what criteria can we determine what is ethically appropriate laughter?

The criterion is obtained by establishing a direct connection between health and ethics. I argue that an ethical life is a healthy life. In other words, the same biological rationality that aspires to health and survival in comfortable conditions also transmits to us demands of ethical behavior that lead us to try to approach this aspiration.

This affirmation is what I argue throughout my work, starting precisely from the messages that our biology seems to transmit through smiles and laughter.

In the following, I will show some aspects of the ethics of humor as applied ethics.

9 The Ethics of Humor as Applied Ethics: The Limits of Humor

A topic that has aroused much media interest is that of clarifying what the limits of humor are, that is, determining when laughing can be considered ethically incorrect. One book that has analyzed humor from this perspective is edited by Sharon Lockyer and Michael Pickering, entitled *Beyond a Joke. The Limits of Humor* [38].

A few years ago, all the major newspapers in Western Europe referred to an offensive remark by Silvio Berlusconi, when he began his term as President of the European Parliament in Strasbourg. He referred to the German politician Martin Schulz saying that he should play the role of a guard in a film about Nazi concentration camps. Most of those present were dismayed, but Berlusconi laughed loudly after hurling his insult.

When he learned of the negative press comments about his words, he justified himself by saying that his comment had simply been a joke and nothing more. As Pickering and Lockyer say, this justification was based on "the commonplace notion that a joke is *sui generis* and should not be registered within the same schema of understanding as serious discourse" ([38]: 1). But our authors think there are ethical limits to what we can say with a joke. This one, in particular, was seriously offensive. The line had been crossed.

Pickering and Lockyer, in the aforementioned book, attempt to answer questions such as the following: How do we measure the aesthetic values of humor against the ethics of avoiding harm? There may be cases in which humor should be cruel, even brutal, and designed to offend or harm its targets, but are these cases exceptional and justifiable only in exceptional circumstances? If not, can we draw a clear line between comic attacks on those with enormous power and those who are relatively weak and undefended? And where does satirical ridicule end and bullying mockery begin? Is there a distinction between questioning and breaking social conventions, or between humor that challenges taboos and humor that is intended to be seriously offensive? Who gets to define what is acceptable and what is not, if we all have an equal right to laugh at whatever we want? Can there be only relative standards of humor in an increasingly fragmented society, with different groups and communities with quite different norms, principles and standards? How do we set an ethical limit to humor without it being considered "censorship"?

They detect a "conflict between those concerned to protect freedom of speech and those concerned to protect minority, oppressed or previously persecuted groups from the public expression of bigotry, misprision, abusive stereotyping, discrimination and hatred" ([38]: 8).

But the ethics of humor, as applied ethics, has much more to contribute. Its main contribution lies in improving each of the social spheres (professions, organizations, and citizenship in general) by developing ethical humor in them. Here I will give three examples of areas in which the ethics of humor is applied: health, business, and education.

10 Applying the Ethics of Humor to Health Care

Numerous books have been published internationally applying studies on humor to the field of health, and one of the first concerns of their authors usually lies in defining a concept of humor that is really healthy or, as they also say, a "therapeutic humor."

Garanto Alós defines humor in this way:

> [it] is the state of mind more or less persistent and stable, that bathes balanced feelings, emotions, moods, or bodily states, arising from the contact of the individual (corporeality and psyche) with the environment, and that enables the individual, taking the appropriate distance, to critically relativize all kinds of affective experiences that are polarized, either towards euphoric situations, either towards depressive situations. [24]

Defined in this way, humor is an indicator of a person's health. But there are types of humor that can harm others, such as aggressive humor, which aims to ridicule others in order to cause them harm, or self-destructive humor. Therefore, I think it is necessary to distinguish between good and bad humor. The former is truly therapeutic.

One of the most interesting books in Spanish in this line is the one by Begoña Carbelo entitled *El humor en la relación con el paciente* (Humor in relationship with the patient) [12]. There the author tells us that therapeutic humor "has to facilitate communication and empathy between the patient and the professional, and create the appropriate climate so that feelings, doubts, and fears can be expressed" ([12]: 31).

The WHO in its annual report of October 2001, dedicated for the first time to mental health, pointed out that in the near future mental disorders will be the second most frequent condition in all Western countries; but governments have no mental wellbeing programs. Stress is known to cause one in four people to suffer from mental illness, and it is a state of excessive tension that is very common in this competitive society. A sense of humor, on the other hand, can relativize depressive and distressing situations.

Authors who write about the health benefits of humor often conclude that health professionals are challenged to lose the fear of smiling and laughing with the patient, as humor makes the relationship more human, and also contributes to greater flexibility and openness.

Begoña Carbelo tells us that, in health, the humor that works is white humor or, what is the same, "a humor capable of providing an environment brimming with healthy laughter that helps to manage stress and achieve adaptation, without offending, without belittling" ([12]: 39).

So, some of the benefits of therapeutic humor or "white humor," which I call ethical humor, are as follows:

1. Creates an appropriate climate in the relationship with the patient.
2. Relativizes depressive and distressing situations.
3. Humanizes the care relationship.

Other benefits of humor that are often mentioned are the following:

1. It provokes feelings of control in anxious situations.
2. It calms in the preoperative period and reduces recovery time.
3. Stimulates the immune system and fights stress.
4. Stimulates the secretion of endorphins, which have an analgesic effect.

Because of these positive effects, laughter therapy has been developed as a new therapy to help patients improve their health by provoking great laughter.

However, we must not forget that not all laughter is healthy. Some forms of laughter are harmful to oneself and to others: self-denigrating laughter and laughter that attacks other people. But do we realize what kind of laughter we usually use?

The ethics of humor helps us to distinguish between healthy and unhealthy humor [59].

The need to distinguish between good and bad humor arose when the first studies showing negative consequences of humor appeared. Carver, Scheier, and Weintraub conducted a study in 1989 to assess the consequences of mood in 528 doctors and nurses in twenty hospitals in northern Italy, who, on the one hand, completed measures of their amount of humor and, on the other hand, measures of anxiety, depression, and emotional exhaustion [13].

Surprisingly, analysis of the data revealed that higher levels of coping humor were associated with greater emotional exhaustion and feelings of depersonalization. Why did their humor have these negative consequences? It wasn't easy to know, because this study measured only the amount of humor, not the type of humor used.

Joan Sayre, in an attempt to answer that question, conducted research in 2001 on the use of humor by staff in a psychiatric unit, but she did establish two categories [51]:

1. A rather benevolent "funny" type (incongruous witticisms, humor about oneself).
2. Another, more aggressive "sarcastic" type (belittling, malicious humor).

Sarcastic humor was more common than funny humor among the staff, and most of the humor was devoted to making fun of patient behavior out of earshot of the patients.

The author suggested that the more aggressive forms could have harmful therapeutic consequences, as well as being morally wrong. Subsequently, Rod A. Martin and his collaborators created a scale that measured four styles of humor, very valuable for ethical reflection [40]. These styles of humor were as follows:

1. Affiliative Humor.
2. Self-enhancing Humor.
3. Aggressive Humor.
4. Self-defeating Humor.

The first two would be healthy (for others or for oneself) and the second two would be unhealthy (for others or for oneself).

Personally, I consider it very important that we equip ourselves with instruments that adequately measure the styles of humor, both of healthcare staff and patients, something that we have not yet developed adequately. I argue that the style of humor is not something fixed in us forever, as we can change it through the right education. If we educate in the right humor, we will be laying the foundations for healthier citizens now and in the future.

11 Applying the Ethics of Humor to the Company

Numerous books in recent years have applied humor to the business world, for example, the following: E. Blumenfeld and L. Alpern, *Humor at Work* [9]; M. Kushner, *The Light Touch. How to Use Humor for Business Success* [35]; J. Morreall, *Humor Works* [42]; T. Paulson, *Making Humor Work* [46]; M. Weinstein, *Managing to Have Fun* [64]; M Weinstein and L. Barber, *Work Like your Dog. Fifty Ways to Work Less, Play More, and Earn More* [65].

Here, to analyze the relationship between humor and the company, I will rely mainly on the book by Eduardo Jáuregui and Jesús Damián Fernández, *Alta diversión. Los beneficios del humor en el trabajo* [31], and, in part, also on Leslie Yerkes, *Fun Works. Creating Places Where People Love to Work* [66], which I will complement with the reflection of other authors on the philosophy and ethics of humor.

Eduardo Jáuregui and Jesús Damián Fernández tell us that, in business, seriousness is the order of the day. The aim is to be taken seriously by the groups affected. For this reason, in communications we use an aseptic, formal tone, full of technicalities. The dress code demands sobriety and often discomfort. Working is supposed to be unpleasant.

But the distinction between work and play is a recent cultural product of human history. It comes from America's early settlers, the Calvinist Puritans, who believed that it was necessary to work hard and avoid the pleasures of this life in order to receive a divine reward in the next.

Work was not always such a solemn activity and some companies are betting on recovering humor at work.

Numerous scientific studies are showing us the substantial benefits of introducing humor in the company. Large multinationals such as Google, Southwest Airlines, and Ben and Jerry's are outstanding examples that have managed to introduce fun experiences for their workers with a significant improvement in their bottom line.

Authors such as Jáuregui and Fernández have been pointing out that a positive mood has the following benefits at work:

1. Reduces stress, and enhances the employee's health and capabilities.
2. Strengthens individual and collective motivation.
3. Attracts and retains the most valuable human resources.
4. Stimulates innovation.
5. Optimizes internal communication.
6. Promotes learning.
7. Improves cohesion of human teams.
8. Enhances the persuasive impact of sales messages.
9. Strengthens customer relationships.
10. Creates a more pleasant and human environment to work in.

For my part, I support these initiatives and I also base these beneficial effects of "positive humor" on what I call ethical humor—a humor that is balanced with oneself and with others.

But, as a consequence of the profitability of humor, there are already a good number of companies that have decided to introduce it in their human teams. Unfortunately, some of them are doing it without a proper ethical orientation and catastrophic consequences.

Thus, employees are being offered what in the business world is already beginning to be called "organized fun," which includes innovations such as "backwards day," where employees are asked to dress upside down, or "silly hat day." But who organizes the fun and for what purpose? Such forced fun can become a form of enforced happiness and a way of intruding into employees' private lives in the employer's interest of increasing profits.

Simon Critchley, in his book *On Humour* [16], tells us about his experience in a hotel in Atlanta, where the employees played games so that the management wouldn't see them as bad. They did not want to give the impression of being bad players or exhausted people. No one had offered them an alternative.

In reality, when employees feel subjugated by authoritarian bosses, they like to be able to joke about them. On many occasions, those who work in a factory or office create, or hear from colleagues, jokes, stories or even songs, making fun of the most imposing actions of the boss. Employees often laugh at the arrogance, the bad decisions, the authoritarianism, etc. of the boss. Laughter seems to act as an escape valve—a survival mode to maintain mental health in stressful situations. But the ethics of humor cannot limit itself to proving the existence of these jokes. It aims to understand why these jokes are occurring, and what actions by bosses are generating them. Jokes about bosses can be interpreted as real guidelines for bosses to reflect on their attitude. The ultimate goal of ethics cannot be to keep laughing at each other, but to interpret the message of laughter to ethically transform the actions of its members and the values of the company as a whole.

12 Applying the Ethics of Humor to Education

An important line of work is that which applies studies on humor to the field of education, in which interesting books have been published [20, 21, 25, 26, 62, 63].

Rod A. Martin has described where the relationship between humor and education is heading [39]. According to this author, pedagogical trends in recent decades promote a more relaxed learning environment with an emphasis on "making learning more fun." The prevailing philosophy of education advocates is that students are much more likely to be motivated to learn and retain information if they are happy and having fun than if they feel anxious and threatened [45].

Most works define pedagogy as "the art and science of education." Jesús Damián Fernández Solís coined the term "Pedagogy of humor" to refer to "that discipline which, having education as its object, incorporates humor as a processual and methodological key to its interventions" ([19]: 82).

Humor applied to the educational field is much more than the art of telling jokes. It awakens a style of teaching and learning. It provides methodologies, techniques, and educational resources.

Fernández Solís believes that education for students should not follow the tortuous paths of boredom, tedium, and drudgery. Learning is more than the accumulation of content without analysis or reflection.

The sense of humor applied to formal and non-formal education presents a new disposition to react to formative processes ([5]: 19–25). It allows us to assemble and transform daily work into a correct attitude towards life.

It is important to consider humor as an irreplaceable component of education. Jesús Damián Fernández Solís tells us that applying a sense of humor in education helps us to overcome frustrations, to relativize failures, to overcome disappointments, to banish moments of apathy and discouragement in order to establish good communication with ourselves and with the environment where we carry out our educational work ([19]: 65–90).

Based on the purported cognitive, emotional, social, and physiological benefits of humor, some educators have even suggested that one of the goals of education should be to facilitate the development of a good sense of humor in students [8, 41].

Didactics is "the science and art of teaching." A sense of humor is a great ally in didactics [20], for several reasons:

1. Motivates learners to learn.
2. It allows for a relaxed and pleasant atmosphere.
3. It favors the empathy of the educator.
4. It enhances the sympathy between educator/educatee.
5. Facilitates new work tools.
6. Exercises creative teaching skills and abilities.

Many educators have recommended that teachers incorporate humor in the classroom with techniques such as the following:

1. Incorporate anecdotes, illustrations, and humorous examples into the lessons.
2. Put funny pictures and phrases on the classroom walls.
3. Encourage students to produce humor frequently.

Several studies have treated humor as a very useful and effective teaching tool [15, 61, 63]. But they have also warned that on many occasions the humor used by teachers can be negative (insulting jokes, making fun of a student for a mistake, etc.).

As Bryant and Zillmann observed, success in teaching with humor depends on the right kind of humor, under the right conditions, at the right time, and with appropriately motivated and receptive students ([10]: 74).

It has been suggested [7, 14, 17, 67] that good humor, or what I call ethical humor, brings the following benefits in classrooms:

1. Reduces tension, stress, anxiety and boredom.
2. Improves the student–teacher relationship.
3. Makes the class less intimidating for students.
4. Makes learning fun, creating positive attitudes towards it.
5. Stimulates interest and attention to educational messages.
6. Increases comprehension, cognitive retention and performance.
7. Promotes creativity and fun thinking.

The use of humor has been seen as an especially valuable tool in educating students about sensitive and anxiety-provoking topics such as death and suicide [32].

13 Conclusions

1. Humor is the ability to perceive something as comical, which activates the emotion of mirth, which is expressed through smiling or laughing.
2. Throughout the history of philosophy, there have been at least three important theories about the causes of the comic: the theory of superiority, the theory of incongruity, and the theory of tension release.
3. Reason tells us, at times, that it is not right to laugh, limiting the emotion of mirth, or at least the expression of it by laughter.
4. There can be an ethics of laughter because laughter is a reflection of a person's character. Where the attitude presupposed by a joke is morally questionable, then telling the joke or laughing at it is also morally questionable.
5. Laughter teaches us a life plan immune to ridicule.
6. We can distinguish two types of *humor ethics*: the *negative*, which would defend the importance of limiting humor because it promotes vices; and the *positive*, which describes the moral virtues that can be fostered through humor.
7. Between buffoonery and complete seriousness, there is a middle ground in which lies the virtue that leads to happiness, by practicing humor at the right time and place, and to the right degree.
8. Ethical humor promotes the following ethical virtues: patience, tolerance, kindness, humility, perseverance, and courage.
9. The *ethics of humor*, in my proposal, is both an ethical theory and an applied ethics.
10. I argue that, at present, in philosophy we are moving from the paradigm of language to what I call the paradigm of biology.
11. Our biology contains a rationality, which I call biological rationality that pursues the metaphysical goal of health and survival in comfortable conditions. I argue that humor is one of the means through which this biological rationality is expressed.
12. The *discursive rationality of the ethics of humor* tries to critically interpret the messages that our biological rationality transmits to us through smiles and laughter.

13. For the interpretation of the messages of biology, the ethics of humor resorts to a method of its own which I have called the "clinical-ethical method."
14. I propose *a new ethical theory based on humor*, that locates in humor the Archimedean point from which we can advance our own understanding as moral subjects, obtain reasons for our moral performance, and show ways to improve ethically both as individuals and as a society as a whole.
15. I believe that it is important to educate in *ethical humor*, that is, humor that knows how to laugh at the right time, in the right way and to the right degree, and for the right reasons.
16. I believe that an ethical life is a healthy life.
17. In analyzing the limits of humor, we detected a conflict between those concerned with protecting freedom of expression and those concerned with protecting minority, oppressed, or previously persecuted groups from public expression of bigotry, whitewashing, abusive stereotyping, discrimination, and hatred.
18. There are types of humor that can harm others. Therefore, I consider it necessary to distinguish between good and bad humor. The former is truly therapeutic, from a health point of view.
19. Therapeutic humor has many health benefits, including the following: it creates an appropriate climate in the relationship with the patient; relativizes depressive and distressing situations; and humanizes the care relationship.
20. The style of humor is not something fixed in us forever, we can change it through the right education. If we educate in the right humor, we will be laying the foundation for healthier citizens now and in the future.
21. Ethical humor has many benefits at work, including the following: it reduces stress and enhances employee health and capabilities; strengthens individual and collective motivation; attracts and retains the most valuable human resources; stimulates innovation; and optimizes internal communication.
22. Students are much more likely to be motivated to learn and retain information if they are happy and having fun than if they feel anxious and threatened.
23. One of the aims of education should be to facilitate the development of a good sense of humor in students.
24. Ethical humor has many benefits in education, including the following: it reduces tension, stress, anxiety, and boredom; improves the student–teacher relationship; stimulates interest and attention to educational messages; increases comprehension, cognitive retention, and performance; promotes creativity and fun thinking.
25. *Ethical humor*, which should be encouraged in all spheres of society, is humor that is balanced towards oneself and others.

References

1. Apel K-O (1973) Transformation der Philosophie, vol 2. Suhrkamp, Frankfurt am Main
2. Aquinas T (1972) Summa Theologiae, vol 44. Well-tempered passion. (Trans. T. Gilby). McGraw-Hill, New York
3. Aristotle (1941) In: McKeon R (ed) The basic works of Aristotle. Random House, New York
4. Basu S (1999) 'Woe unto you that laugh now!': humor and toleration in Overton and Shaftesbury. In: Laursen, JC (ed) Religious toleration. "The varieties of rites" from Cyrus to Defoe. St. Martin's, New York, pp 147–172
5. Benigno J (1996) Humor y seriedad. En la vida corriente. Eunsa, Navarra
6. Bergson H (1940) Le rire. Essai sur la signification du comique. Presses Universitaires de France, Paris
7. Berk RA, Nanda JP (1998) Effects of jocular instructional methods on attitudes, anxiety and achievement in statistics courses. Humor Int J Humor Res 11(4):383–409
8. Bernstein DK (1986) The development of humor. Implications for assessment and intervention. Top Lang Disord 4(4):65–71
9. Blumenfeld E, Alpern L (1994) Humor at work. Peachtree Publishers, Atlanta
10. Bryant J, Zillmann D (1989) Using humor to promote learning in the classroom. In: McGhee PE (ed) Humor in children's development. A guide to practical applications. Haworth Press, New York, pp 49–78
11. Buckley FH (2005) The morality of laughter. The University of Michigan Press, Ann Arbor
12. Carbelo B (2008) El humor en la relación con el paciente. Una guía para profesionales de la salud. Masson, Barcelona
13. Carver CS, Scheier MF, Weintraub JK (1989) Assessing coping strategies: a theoretically based approach. J Pers Soc Psychol 56(2):267–283
14. Codina Felip MJ (2015) Neuroeducación en virtudes cordiales. Cómo reconciliar lo que decimos con lo que hacemos. Octaedro, Barcelona
15. Cornett CE (1986) Learning through laughter. Humor in the classroom. PhiDelta Kappa Educational Foundation, Bloomington-IN
16. Critchley S (2002) On humour. Routledge, London
17. Davies AP, Apter MJ (1980) Humor and its effect on learning in children. In: McGhee PE, Chapman AJ (eds) Children's humour. Wiley, Chichester, pp 237–253
18. De Sousa R (1987) When is it wrong to laugh? In Morreall J (ed) The philosophy of laughter and humor. State University of New York Press, Albany-NY (That article was published that same year as chapter 11 of his book entitled *The Rationality of Emotion*, Massachusetts Institute of Technology, 1987, 5th reprint 1997. Cited as compiled by J. Morreall)
19. Fernández Solís JD (2002) Pedagogía del humor. In: Rodríguez Idígoras Á (ed) El valor terapéutico del humor. Desclée De Brouwer, Bilbao
20. Fernández Solís JD, Francia A (1995) Animar con humor. Aprender riendo, gozar educando. CCS, Madrid
21. Francia A, Fernández Solís JD (2009) Educar con humor. Aljibe, Málaga
22. Freud S (1963) Jokes and their relation to the unconscious. Standard Edition of the complete psychological works of Sigmund Freud, vol 5 (Trans. J. Strachey). W.W. Norton, New York (Original publication in 1905)
23. Freud S (1987) Humour. In: Morreal J (ed) The philosophy of laughter and humor (Trans. J. Rivière). State University of New York Press, Albany, pp 111–116 (Original publication in 1927)
24. Garanto Alós J (1983) Psicología del humor. Herder, Barcelona
25. García Cerrada M, Fernández Solís JD (2010) El valor pedagógico del humor en la educación social. Desclée de Brouwer, Bilbao
26. García Larrauri B (2010) Claves para aprender en un ambiente positivo y divertido. Todo rueda mejor si se engrasa con humor. Pirámide, Madrid

27. Hobbes T (1839) Leviathan. In: Molesworth W (ed) The English works of Thomas Hobbes, vol 3. Bohn, London
28. Höffding H (2007) Humor als Lebensgefühl (Der grosse Humor). Eine psychologische Studie. Dr. Müller e. K. und Lizenzgeber, Saarbrücken
29. Holy Bible (2011) New international version. NIV. Biblica, Inc
30. Hösle V (2001) Woody Allen. Versuch über das Komische. Verlag C.H. Beck, Munich
31. Jáuregui E, Fernández Solís JD (2008) Alta diversión. Los beneficios del humor en el trabajo. Alienta, Barcelona
32. Johnson HA (1990) Humor as an innovative method for teaching sensitive topics. Educ Gerontol 16(6):547–559
33. Kant I (1892) Critique of judgment (Trans. J.H. Bernard). Macmillan, London
34. Kierkegaard S (2009) In: Hannay A (ed) Concluding unscientific postscript. Cambridge University Press, Cambridge
35. Kushner M (1990) The light touch. How to use humor for business success. Simon and Schuster, New York
36. Lipman S (1991) Laughter in hell. The use of humor during the Holocaust. Jason Aronson, Northvale-NJ
37. Lipps T (2013) Komik und Humor. Eine Psychologische-Ästhetische Untersuchung. Tredition, Hamburg (original publication 1898)
38. Lockyer S, Pickering M (eds) (2009) Beyond a Joke. The Limits of Humour. Palgrave McMillian, New York (first published 2005)
39. Martin RA (2007) The psychology of humor: an integrative approach. Elsevier, Amsterdam
40. Martin RA, Puhlik-Doris P, Larsen G, Gray J, Weir K (2003) Individual differences in uses of humor and their relation to psychological well-being: development of the humor styles questionnaire. J Res Pers 37(1):48–75
41. Masselos G (2003) 'When I play funny it makes me laugh'. Implications for early childhood educators in developing humor through play. In: Lytle DE (ed) Play and educational theory and practice. Praeger Publishers-Greenwood Publishing, Westport-CT
42. Morreall J (1997) Humor works. Human Resource Development Press, Amherst-MA
43. Morreall J (2009) Comic relief. A comprehensive philosophy of humor. Wiley-Blackwell, Malden, MA
44. Nietzsche F (2013) Also sprach Zarathustra. Kröner, Stuttgart
45. Oppliger PA (2003) Humor and learning. In: Bryant J, Roskos-Ewoldsen D, Cantor JR (eds) Communication and emotion. Essays in honor of Dolf Zillmann. Lawrence Erlbaum Associates, Mahwah-NY
46. Paulson T (1989) Making humor work. Crisp Publishing, Los Altos-CA
47. Pirandello L (2007) L'Umorismo. Essenza, caratteri e materia dell'umorismo. El humorismo (Esencia, carácter y materia del humorismo) (1908). Bilingual edition. San Lorenzo del Escorial: Langre (in this Italian-Spanish bilingual edition, the second part of the essay is reproduced)
48. Plato (2005). Republic. In: Hamilton E, Cairs H (eds) The collected dialogues of Plato. Princeton University Press, Princeton
49. Plato (2005). "Philebus." In: Hamilton E, Cairs H (eds) The collected dialogues of Plato. Princeton University Press, Princeton
50. Potter VR (1970) Bioethics, science of survival. Biol Med 14:127–153
51. Sayre J (2001) The use of aberrant medical humor by psychiatric unit staff. Issues Ment Health Nurs 22(7):669–689
52. Shaftesbury AAC, Earl (2008) A letter concerning enthusiasm. Sensus communis; an essay on the freedom of wit and humour. Soliloquy, or Advice to an Author (1). HardPress Publishing (work originally published in 1711)
53. Siurana JC (2012) Filosofía del humor: estado de la cuestión. Diálogo Filosófico 82:4–33
54. Siurana JC (2012) Fundamentos para una ética del humor. In: Casaban Moya E (ed) XIX Congrés Valencià de Filosofia. Societat de Filosofia del País Valencià, Valencia, pp 473–484

55. Siurana JC (2012) Ética del humor en el ámbito empresarial para tiempos de crisis económica. In: Ballester R et al (coords.) Actas del XX Congreso EBEN-España. La imagen de la empresa en el siglo XXI. ÉTNOR Foundation, Valencia, pp 491–530
56. Siurana JC (2013) Libertad, justicia y neurociencias en el final de la vida desde la ética del humor. In: López Frías FJ et al (eds) Bioética, Neuroética, Libertad y Justicia. Comares, Granada, pp 845–866
57. Siurana JC (2013) Los rasgos de la ética del humor. Una propuesta a partir de autores contemporáneos. Veritas 29:9–13
58. Siurana JC (2014) Ética del humor y diversidad cultural. Dilemata Int J Appl Ethics 6(15):215–231
59. Siurana JC (2015) Ética del humor. Fundamentos y aplicaciones de una nueva teoría ética. Plaza y Valdés, Madrid
60. Siurana JC (2015) The capability for ethical humor. In: Nakatogawa K, de Tienda Palop L, Mitsuke Y, Fukayama Y (eds) Discussing capabilities, emotions and values: a cross-cultural perspective. Key Word, Sapporo-Japan, pp 65–88
61. Struthers A (2003) No laughing! Playing with humor in the classroom. In: Klein AJ (ed) Humor in children's lives. A guidebook for practitioners. Praeger, Westport
62. Tamblyn D, Weiss S (2000) The big book of humorous training games. McGraw-Hill, New York
63. Tamblyn D (2003) Laugh and learn. 95 ways to use humor for more effective teaching and training. Amcom, New York
64. Weinstein M (1997) Managing to have fun. Simon and Schuster, New York
65. Weinstein M, Barber L (1999) Work like your dog. Fifty ways to work less, play more, and earn more. Villard, New York
66. Yerkes L (2001) Fun works. Creating places where people love to work. Berrett-Koehler, Oakland
67. Ziegler V, Boardman G, Thomas MD (1985) Humor, leadership, and school climate. Clear House 58:346–348

SPRINGER NATURE

GPSR Compliance

The European Union's (EU) General Product Safety Regulation (GPSR) is a set of rules that requires consumer products to be safe and our obligations to ensure this.

If you have any concerns about our products, you can contact us on ProductSafety@springernature.com

In case Publisher is established outside the EU, the EU authorized representative is:

Springer Nature Customer Service Center GmbH
Europaplatz 3
69115 Heidelberg, Germany

The manufacturer's authorised representative in the EU is Springer Nature Customer Service Centre GmbH, Europaplatz 3, 69115 Heidelberg, Germany. If you have any concerns regarding our products, please contact ProductSafety@springernature.com

Printed and bound by CPI Group (UK) Ltd, Croydon, CR0 4YY
30/09/2026
02225471-0017